Advances in Comminution

Advances in Comminution

Editor

Manoj Gupta

scitus
academics

Advances in Comminution

Edited by **Manoj Gupta**

Printed in 2017

ISBN: 978-1-68117-471-6

Library of Congress Control Number: 2015936589

© 2016 by

SCITUS Academics LLC,
616, Corporate Way, Suite 2, 4766,
Valley Cottage, NY 10989

www.scitusacademics.com

Contents

Preface

Comminution is the reduction of solid materials from one average particle size to a smaller average particle size, by crushing, grinding, cutting, vibrating, or other processes. In geology, it occurs naturally during faulting in the upper part of the Earth's crust. In industry, it is an important unit operation in mineral processing, ceramics, electronics, and other fields, accomplished with many types of mill. In dentistry, it is the result of mastication of food. In general medicine, it is one of the most traumatic forms of bone fracture. Within industrial uses, the purpose of comminution is to reduce the size and to increase the surface area of solids. It is also used to free useful materials from matrix materials in which they are embedded, and to concentrate minerals.

Editor

Estimation of Elasticity of Porous Rock Based on Mineral Composition and Microstructure

Zaobao Liu[1,2], Jianfu Shao[1,2], Weiya Xu[1],
and Chong Shi[1]

[1]Geotechnical Research Institute, Hohai University, Nanjing 210098, China

[2]Laboratory of Mechanics of Lille, University of Lille 1, 59655 Villeneuve d'Ascq, France

ABSTRACT

Estimation of elastic parameters of porous rock like the compressibility of sandstone is scientifically important and yet an open issue. This study illustrates the estimation of the elastic compressibility of sandstone (ECS) based on the assumption that the ECS is determined closely by the mineral composition and microstructures. In this study, 37 samples are collected to evaluate the estimations of the ECS obtained by different methods. The regression analysis is first implemented using the 37 samples. The results show that ECS exhibits linear relations with the rock minerals, pores, and applied compressive stress. Then the support vector machine (SVM) optimized by the particle swarm optimization algorithm (PSO) is examined to generate estimations of the ECS based on the mineral composition and microstructures. The SVM is trained with 30 samples to search for optimal parameters using the PSO, and thus the estimation model is established. Afterwards, this model is validated to give predictions of the left 7 samples. By comparison with the regression methods, the proposed strategy, that is, the PSO optimized SVM, performs much better on the training samples and shows a good capability in generating estimations of the ECS of the 7 testing samples based on the mineral composition and microstructures.

INTRODUCTION

The identification of elastic parameters of porous rock is one of the major problems in rock mechanics and so far an open issue. The elastic properties of porous rock show apparent variability from one project to another in practice. In the formulation of suitable constitutive models, the material parameters should be determined first from relevant experimental data in order to describe the mechanical behaviors of rock materials under different engineering contexts [1]. Classical deterministic approaches [2–4] have been firstly used to identify the physical properties of rock materials, generally based on laboratory experiments [5–9] and in situ tests [10,11]. However, these tests sometimes are difficult to be realized and may involve heavy costs. In this way, estimation of the coefficients related to the physical properties has also drawn much attention for feasibility and easiness in practice.

Approaching this issue, many techniques have been proposed for the estimation of elastic parameters of porous rock material [4, 12–18]. In these techniques, the elastic parameters are thought to be closely related to some other indexes that are easily to be determined. However, these conventional methods, for example, the empirical equations, have very poor generalization ability in estimation. This is doomed due to the insufficiency of these methods to account for the uncertain relationships between the elastic parameters and the related indexes such as the rock minerals and microstructures.

In order to better take into account the uncertainties in the determination of elastic parameters of rock materials, various soft computing methods have been introduced to approach this problem in the past decades [19]. These methods provide a new way for the description of elastic parameters of rock materials since with such approaches it becomes possible to learn some disciplines among the related rock parameters from the relevant data obtained. In this manner, if similar positive results can be found, expensive experimental identification procedures can be avoided. Towards this issue, some valuable results have been obtained in some previous works by using the neural networks and regression techniques [20–25]. In these works, both the laboratory tests and field measurements have been used for estimating the elastic parameters. However, there is a common shortcoming for both the field tests and the laboratory tests. They cannot consider all the physical and mineral parameters such as the mineral composition, the particle size, and the distribution of voids. And what is more, as mentioned above, they are expensive to be realized in some cases. For the simple neural networks, they have some shortcomings all the same, such as the local solution, weak generalization ability, and high computational expense. In this way, more effective methods are still in need to approach the estimation of the elastic parameters of the porous rock materials.

In this paper, we illustrate estimation of the elastic parameters of sandstone (the one common material in geotechnical engineering and earth science) according to their mineral compositions and microstructural properties using the SVM model [26, 27]. We apply the particle swarm optimization algorithm (PSO) to optimize the SVM model parameters which have been proved to have a significant effect on the model performance [28]. We demonstrate the applicability and reliability of this method noted as the PSO-SVM for estimation of rock elastic compressibility based on the experimental data of the rock mineral compositions and microstructure features as well as the loading pressure. The other elastic parameters of porous rock can be estimated using the strategy in a similar way and are thus not discussed in this paper.

MATERIAL AND METHOD

Rock Elastic Compressibility and Experimental Data

Compressibility is a measure of the relative volume change of material as a response to a stress (or hydrostatic pressure) change under certain conditions. The measurement of rock compressibility is accomplished through measuring the change of pore volume versus pore pressure. The rock compressibility usually has an unreasonable deviation from its true value for the reason that the measurement is invisibly affected by many uncertain parameters.

Rock compressibility is the volume shrunken feature of rocks under pressure, and is reflected by the compressibility coefficients. The compressibility coefficients of the rock are closely related to the ambient pressure, as well as the fracture distribution, the mineral and its proportion, density, and void ratio. The beginning work on rock compressibility that is widely used today is done by Hall [29] in 1953. He developed a graph of the rock compressibility versus porosity by statistical analysis of laboratory experiments, which is called Hall's plot today and simulated by some empirical formula. Then a similar fundamental work has been done by Newman [30]. He obtained a similar trend of rock compressibility with porosity to Hall's plot for both the consolidated sandstone and limestone.

However, Hall's plot, in some cases, shows a logically confusion relation between the compressibility and porosity in rocks. According to Hall's plot, the compressibility decreases as rock porosity increases. Extremely tight rocks have an abnormally high compressibility [31]. In fact, tight rocks are less compressible than loose rocks and should have a smaller compressibility. Moreover, the rock compressibility by Hall's plot is usually larger than that of the reservoir liquids in the normal range of reservoir porosity. Thus, Hall's plot is not sufficient enough in every case. Also, Hall's plot gives the same compressibility value for rocks of different lithology if only they have the same porosity regardless of their different rigidity. That is to say, Hall's plot does not consider the effect of rock lithology and minerals composing the rocks.

Some other discussions on the rock compressibility can also be found in the pieces of literature [32–36].

Experiments have been done on sandstones for discovering the relationships between the compressibility coefficients and their mineral compositions, voids, and other parameters [3]. As stated in the work, the samples used in the experiments are mostly clean quartz arenites, subarkoses, and argillaceous quartz arenites, in which kaolinite is the dominant pore-filling mineral. Total porosity is divided into three types which are the intergranular (equidimensional, size comparable to grains), the connective (tabular or tubular shaped), and the micro (less than a few microns in size) porosity on the basis of point counting SEM images. The micro pores occur within aggregates of clay. The ranges of porosity of samples are total porosity, 5–31% of whole sample; intergranular porosity, 24–76% of total porosity; connective porosity, 4–25% of total porosity; microporosity, 10–63% of total porosity. Empirical equations of calculating rock compressibility are mainly based on the rock porosity and cannot take into account the effects of all the associated parameters.

The experimental results are rearranged as shown in Table 1 where the three coefficients a_A, a_B, and a_C are the elastic linear compressibility measured by gauges settled in three orthogonal directions. In all, 37 samples of sandstone are used in this study and each sample has 11 features. The box graph of each parameter of all the sandstone samples is shown in Figure 1.

Table 1: Compression experimental results of sandstones

Sample number	Mineralogy (vol %)				Average particle size (μm)	Density (g·cm⁻³)	Average void ratio (%)	Pore distribution (%)			Pressure (MPa)	Compressibility coefficients MPa⁻¹ × 10⁶		
	Quartz	Feldspar	Shard clay	Others				Among particle	Pore wall	Micro pore		a_A	a_B	a_C
1	80	0	18	2	65	2.01	24.1	34	23	43	10	36.2	43.6	26.0
2	85	8	7	0	175	1.98	23.8	67	11	22	100	32.7	24.0	29.3
3	70	15	12	3	90	2.02	22.7	35	14	51	30	42.0	41.8	43.0
4	83	10	3	4	220	1.78	30.5	72	4	24	50	43.1	40.8	35.0
5	65	5	28	2	95	2.00	23.5	50	10	40	10	64.8	94.2	105.0
6	65	5	28	2	95	2.00	23.5	50	10	40	50	44.1	55.2	54.9
7	65	5	28	2	95	2.00	23.5	50	10	40	100	42.0	51.4	50.1
8	80	9	4	7	80	2.21	15.6	57	26	17	30	25.5	31.8	45.4
9	80	9	4	7	80	2.21	15.6	57	26	17	50	22.6	26.8	36.5
10	80	9	4	7	80	2.21	15.6	57	26	17	100	20.2	22.9	30.9
11	94	0	1	5	120	2.14	17.8	72	13	15	10	46.1	54.2	64.0
12	94	0	1	5	120	2.14	17.8	72	13	15	30	28.9	26.8	32.5
13	94	0	1	5	120	2.14	17.8	72	13	15	100	24.0	22.8	23.1
14	95	3	2	0	170	2.13	18.1	76	14	10	10	40.1	50.5	34.0
15	95	3	2	0	170	2.13	18.1	76	14	10	100	19.9	21.6	18.5
16	95	0	5	0	130	2.03	21.7	76	10	14	30	32.8	29.0	30.2
17	98	0	2	0	350	2.33	10.6	69	17	14	10	24.3	27.6	34.1
18	98	0	2	0	350	2.33	10.6	69	17	14	50	14.1	15.6	15.8
19	96	1	2	1	115	1.94	24.1	70	14	16	30	33.2	35.9	35.4
20	96	1	2	1	115	1.94	24.1	70	14	16	100	25.0	28.8	26.0
21	90	2	1	7	100	2.18	16.8	60	19	21	10	34.3	26.2	36.8

22	90	2	1	7	100	2.18	16.3	60	19	21	50	21.8	24.0	23.2
23	92	0	8	0	120	2.42	8.5	24	13	63	50	27.5	22.2	18.9
24	92	0	8	0	120	2.42	8.5	24	13	63	30	36.5	28.5	22.7
25	90	4	6	0	145	2.01	23.4	69	11	20	10	43.4	44.8	50.0
26	70	15	12	3	90	2.02	22.7	35	14	51	10	86.0	67.2	86.4
27	83	5	8	4	65	1.91	27.3	68	10	22	10	63.6	63.8	65.9
28	83	5	8	4	65	1.91	27.5	68	10	22	30	48.5	47.2	53.8
29	85	8	7	0	175	1.98	23.8	67	11	22	30	49.4	49.8	55.1
30	83	10	3	4	220	1.78	30.5	72	4	24	10	46.8	46.4	38.4
31	94	0	1	5	120	2.41	17.8	72	13	15	50	26	24.9	27
32	95	0	5	0	130	2.03	21.7	76	10	14	50	31	26.4	27.3
33	80	9	4	7	80	2.21	15.6	57	26	17	10	34.4	48.6	82.9
34	96	1	2	1	115	1.94	24.1	70	14	16	10	47.8	50.5	50.4
35	90	4	6	0	145	2.01	23.4	69	11	20	100	29	24.3	27
36	90	4	6	0	145	2.01	23.4	69	11	20	50	30.5	27.4	30.9
37	75	9	15	1	85	2.96	25.9	40	16	44	30	58.1	57	55.1

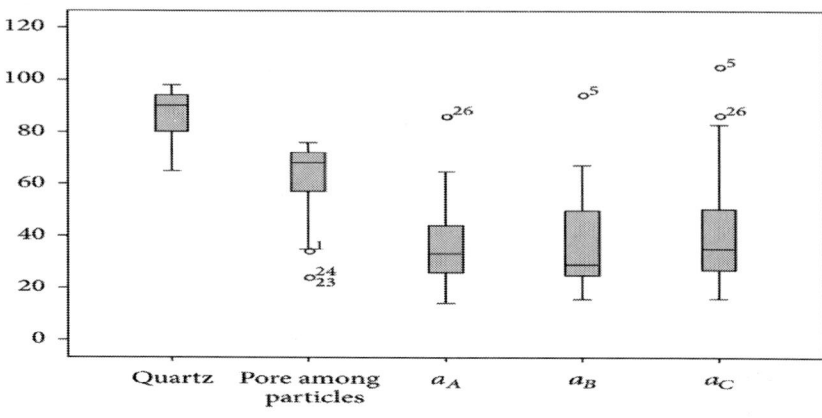

(a)

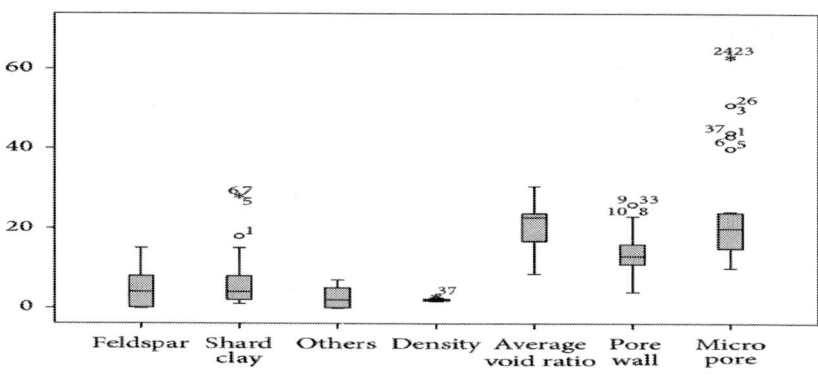

(b)

Figure 1: Box graph of the properties of sandstone experiment data (without unit).

The parameter values of the samples are shown in two subfigures (Figures 1(a) and 1(b)) in order to well illustrate their statistical aspects due to the differences of their value ranges. The horizontal axis of Figure 1 lists the parameter names and the vertical axis denotes the parameter values without units. The small circles (○) and the stars (*)

in Figure 1 exhibit the "outliers" produced by the box graph based on the statistical features of the dataset. The numbers beside the circle and star markers are the sample numbers listed in the first column of Table 1. They are the test parameter values which are not in the statistical range of the box graph. In this manner, these "outliers" do not necessarily mean that they are true outliers and should be removed from analysis of the dataset. Nevertheless, the box graph manifests that some parameters in the dataset have several "outliers" which indicate the moderate quality of statistical consistency of the dataset.

Support Vector Machine

The support vector machines, also known as the support vector networks [26], are supervised learning models with associated learning algorithms which deal with data and recognize patterns and are mainly used for classification and regression analysis.

Given the 30 training samples in Table 1 denoted by $(X, Y) = (x_i, y_i)_{i=1}^n$, here x_i is the ith sample with 11 parameters values (e.g., the mineral type and the density); y_i is the three compressibility coefficients (a_A, a_B, and a_C) of the ith sample; n is the sample number (n=30). The SVMs make a mapping of the samples with a linear regression function

$$y = f(x) = w \cdot \phi(x) + b,$$

(1)

where w is the weight vector, b is bias, and $\varnothing(x)$ is the nonlinear mapping from the input space to output space. The SVMs can efficiently perform nonlinear mappings using what is called the kernel trick, implicitly mapping their inputs into high-dimensional feature spaces.

Suppose that all the samples can be mapped well with a linear function with precision ε. Considering the true mapping errors, the nonnegative slack variables ξ_i and ξ_i^* can be introduced. Thus, the problem can be transformed with the inequalities

$$y_i - w\phi(x_i) - b \le \varepsilon + \xi_i,$$
$$\phi(x_i) w + b - y_i \le \varepsilon + \xi_i^*, \quad i = 1, 2, L, n.$$

(2)

The purpose of SVM training is to minimize the following function:

$$\phi\left(w, \xi_i, \xi_i^*\right) = \frac{1}{2}\left(ww^T\right) + C\sum_{i=1}^{n}\left(\xi_i + \xi_i^*\right),$$

$$(3)$$

where the constant (C>0) is the penalty parameter denoting the punishing level of the samples with errors over. $\in$ Therefore, the problem can be rewritten as

$$\max \quad \left\{\frac{1}{2}\sum_{i=1}^{n}\sum_{j=1}^{n}\left(\alpha_i - \alpha_i^*\right)\left(\alpha_j - \alpha_j^*\right)K\left(x_i, x_j\right)\right.$$

$$\left. -\varepsilon\sum_{i=1}^{n}\left(\alpha_i - \alpha_i^*\right) + \sum_{i=1}^{n}y_i\left(\alpha_i - \alpha_i^*\right)\right\},$$

$$\text{S.t.} \quad \sum_{i=1}^{n}\left(\alpha_i - \alpha_i^*\right) = 0 \quad \left(\alpha_i, \alpha_i^* \in [0, c]\right),$$

$$(4)$$

Where $K(x_i, x_j) = \emptyset(x_i)\emptyset(x_j)$ is the kernel function. There are many commonly used kernel functions, like the multinomial kernel, the sigmoid kernel, and the RBF kernel.

this way, the prediction model of the SVMs can be obtained as follows:

$$f\left(x\right) = \sum_{i=1}^{n}\left(\alpha_i - \alpha_i^*\right)K\left(x_i, x\right) + b.$$

$$(5)$$

Particle Swarm Optimization

The particle swarm optimization (PSO) algorithm is proposed for searching the optimal solution in complex space by the collaboration and competition among particle individuals. It is a population based stochastic optimization technique inspired by social behavior of bird flocking or fish schooling, developed by Eberhart and Kennedy in 1995 [37]. The PSO simulates the foraging behavior of birds. Each solution in the optimizing problem is looked as a bird or a particle in the algorithm in the searching space. The goodness of a particle is evaluated by the value of the fitness function. Each particle keeps track of its coordinates in the problem space which are associated with the best solution (fitness) it has achieved so far. The fitness value is also stored. This value is called P_{best}. Another "best" value that is tracked by the particle swarm optimizer is the best value, obtained so far by any particle in the neighbors of the particle. This location is called L_{best}. The best value is a global best and is called G_{best} when a particle takes all the population as its topological neighbors.

The velocity of the particle i in the n dimensional space is denoted as $V_i = \{V_{i1}, V_{i2},..., V_{in}\}$. The corresponding location is $xi = \{x_{i1}, x_{i2},...,_n\}$; the best solution of the particle is $Pbest = \{p_{i1}, p_{i2},...,p_{in}\}$; the global best solution is $G_{best} = \{g_1, g_2,...,g_n\}$.

The particle swarm optimization concept consists in, at each time step, changing the velocity of (accelerating) each particle toward its P_{bes} and L_{best} locations (local version of PSO). Acceleration is weighted by a random term, with separate random numbers being generated for acceleration toward P_{bes} and L_{best} locations. PSO updates the velocity and location of the particles with the following equation:

$$v_i^{k+1} = wv_i^k + c_1 r\left(\cdot\right)\left(P_{best} - x_i^k\right) + c_2 r\left(\cdot\right)\left(G_{best} - x_i^k\right),$$

$$x_i^{k+1} = x_i^k + v_i^{k+1},$$

(6)

where k is the iteration number; w is the inertia weight; $r(\cdot)$ is a random constant uniformly distributed in the interval$(0, 1)$; c_1 and c_2 are the learning coefficients.

PSO Optimized SVM

The values of the penalty parameter C and the kernel parameters affect directly the model performance in the SVM modeling. Due to the fast and global optimizing features of PSO, it is applied to optimize the parameters in the SVM modeling. The implementation is done in the following steps.

- Initialize the parameters in PSO, including the particle size, the iteration number, the inertia weight w, and the learning coefficients c_1 and c_2.

- Determine the range of the optimized parameters and specify P_{best} and G_{best}.
- Define the fitness function

$$f(x) = \sum_{i=1}^{n} \frac{|u_i - u_i^*|}{u_i},$$

(7)

where u_i is the observed value of the ith sample; u_i^* is the predicted value of SVM; $i = 1, 2, \ldots, n$ is the sample numbers.

- Calculate the fitness value of every particle and compare this value with the P_{best} (the best fitness value of its ever best location L_{best}). If this value is better than P_{best}, then update L_{best} with the new location.
- Compare P_{best} with G_{best}. If P_{best} is better than G_{best}, then renew G_{best} with P_{best}.
- Check whether the fitness value or the iteration number is satisfied with the end condition of the algorithm. If not, update the location and velocity of the particle with (6) or exit and output the results of the coefficients.
- Set up the optimized SVM model for modeling in (5) with the optimized parameters.

The implementation of the estimation is given in Figure 2.

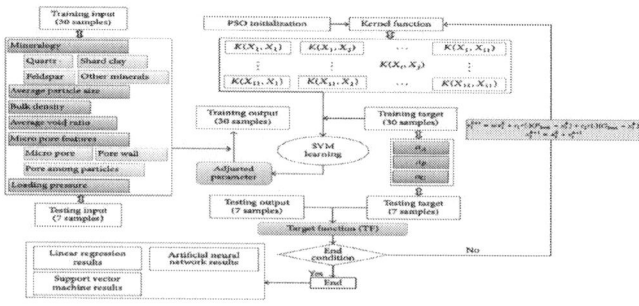

Figure 2: Implementation of estimation of rock elastic compressibility using PSO-SVM.

RESULTS AND DISCUSSIONS

Regression Analysis

We first apply the linear stepwise regression method [38] to analyze the problem. In the stepwise linear regression, the forward method is used to remove the variables in the regression models. The stepping method criteria are the probability (F to remove X_k) > 0.10 to determine whether a parameter X_k is removed. If the inequality does not hold, no variable is removed from the model. If there are no independent variables currently entered in the model or if no entered variable is to be removed, choose X_k such that (F to remove X_k) is maximum. A parameter Xk is entered if (F to remove X_k) < 0.05. If the inequality does not hold, no variable is entered. At each step, all eligible variables are considered for removal and entry [39].

The results of the stepwise regression models are summarized in Table 2 to Table 4, respectively, for the three coefficients a_A, a_B, and a_U. Three predictors are generalized for each coefficient in the regression. It is obvious that the predictor a and predictor b in each model are not physically meaningful at all because the two predictors only retain no more than two potential parameters to explain the compressive coefficients, which is obviously unmeaning. The R square and adjusted R square values are all less than 0.70 for model 3(predictor c) in Table 2 to Table 4. This implies that the compressive coefficients do not have a strict linear relation to the associated parameters. The value "sig. F change" in Tables 2, 3, and 4 shows that the derived regression models are statistically significant (less than 0.005). In short, the stepwise regression can only perform moderately in modeling these sand rock samples.

Table 2: Model summary of stepwise regression for a_A

Model	R	R square	Adjusted R square	Std. error of the estimate	Change statistics				
					R square change	F change	df1	df2	Sig. F change
1	.542[a]	.294	.274	12.53587	.294	14.560	1	35	.001
2	.713[b]	.509	.480	10.60497	.215	14.906	1	34	.000
3	.834[c]	.695	.668	8.47732	.186	20.208	1	33	.000

[a]Predictor: (constant) quartz; [b]predictor: (constant) quartz, pressure; [c]predictor: (constant), quartz, Pressure, and pore wall; [d]dependent variable: a_A.

Table 3: Model summary of stepwise regression for a_B

Model	R	R square	Adjusted R square	Std. error of the estimate	Change statistics				
					R square change	F change	df1	df2	Sig. F change
1	.608[a]	.369	.351	13.44219	.369	20.485	1	35	.000
2	.756[b]	.571	.546	11.24635	.202	16.002	1	34	.000
3	.825[c]	.680	.651	9.85276	.109	11.298	1	33	.002

[a]Predictor: (constant), quartz; [b]predictor: (constant), quartz, pressure; [c]predictor: (constant), quartz, pressure, and pore wall; [d]dependent variable: a_B.

Table 4: Model summary of stepwise regression for a_C

Model	R	R square	Adjusted R square	Std. error of the estimate	Change statistics				
					R square change	F change	df1	df2	Sig. F change
1	.627[a]	.393	.375	15.82344	.393	22.641	1	35	.000
2	.777[b]	.604	.581	12.95912	.212	18.182	1	34	.000
3	.816[c]	.667	.636	12.07530	.062	6.159	1	33	.018

[a]Predictor: (constant), quartz; [b]predictor: (constant), quartz, pressure; [c]predictor: (constant), quartz, pressure, and pore among particles;[d]dependent variable: a_C.

The remaining parameters are the mineral quartz, the pressure, and the pore wall for the coefficients a_A and

a_B in predictor c. For the coefficient a_C, the remaining parameters are quartz, the pressure and the pore among particles in Predictor c. That is to say those parameters related to the rock minerals, applied pressure, and pores are remained exclusively in the stepwise regression results. At this point, the rock compressibility coefficients can be thought to have linear relations with the rock minerals, loading pressure, and pores which can be used to interpret the characteristics of the rock compressibility coefficients.

PSO-SVM Analysis

According to the experimental results, we assume that the minerals of the rock, average size of crystalline particle, interspace distribution, average void ratio, the density of rock, and the pressure of the test specimen are the potential influencing parameters for compressibility coefficients of sand rock materials. We utilize the PSO-SVM to map the relations between the compressibility coefficients and their potential influencing parameters and compare the results with those of the (ANN) [20] and simple SVM models. The first 30 samples listed in Table 1 are used as training samples to establish the models and the last 7 samples are used for testing generalization ability of produced models. The structures of SVM are shown in Figure 2 for the estimation of elastic compressibility of sandstone.

The predicted rock compressibility coefficients of the tested samples are shown in Table 5 for all the introduced techniques. Based on these results, the predictive performance of these approaches is illustrated in Figure 3 to Figure 5 for a_A, a_B, and a_C, respectively. The linear trend lines in the figures show the correlation between the observed value and predicted value. The results of different approaches are shown with different markers and different colors. The R^2 values imply directly the predictive performances. The larger the R^2 value is, the better the corresponding approach performs. It is obvious that the PSO-SVM approach performs best among these techniques regarding the prediction of these three coefficients. Also, the R^2 value in Figure 3 is much smaller than that in Figures 4 and 5, which indicates that these approaches cannot generate as good results of a_A as those of a_B or a_C.

Table 5: Measured and predicted rock compressibility coefficients

Sample number	Measured			ANN prediction			SVM prediction			PSO-SVM prediction		
	a_A	a_B	a_C	a_A	a_B	a_C	a_A	a_B	a_C	a_A	a_B	a_C
1	26.0	24.9	27.0	25.0	25.2	24.3	23.2	23.9	24.8	27.1	24.6	28.5
2	31.0	26.4	27.3	30.0	29.4	25.6	26.8	23.4	25.0	32.1	27.9	28.0
3	34.4	48.6	82.9	59.9	65.1	87.0	30.5	34.1	49.8	42.2	54.4	79.6
4	47.8	50.5	50.4	41.9	39.7	43.9	41.4	43.8	43.8	41.9	55.2	57.7
5	29.0	24.3	27.0	28.1	26.8	24.1	27.5	22.8	25.6	30.1	26.1	26.5
6	30.5	27.4	30.9	34.8	33.0	32.3	33.3	29.6	35.6	31.2	31.5	32.1
7	58.1	57	55.1	54.9	61.6	69.9	40.2	43.8	43	55.1	60.6	62.9

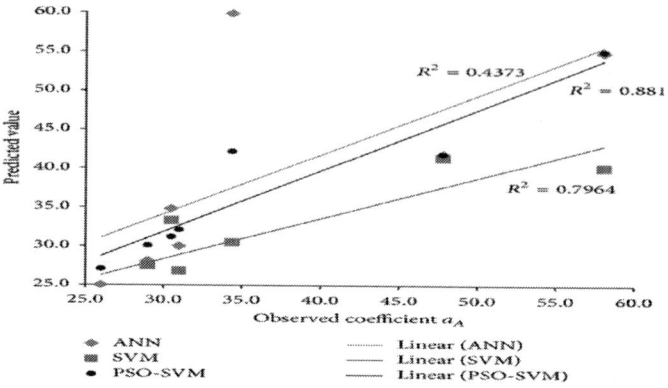

Figure 3: Predictive performance of a_A.

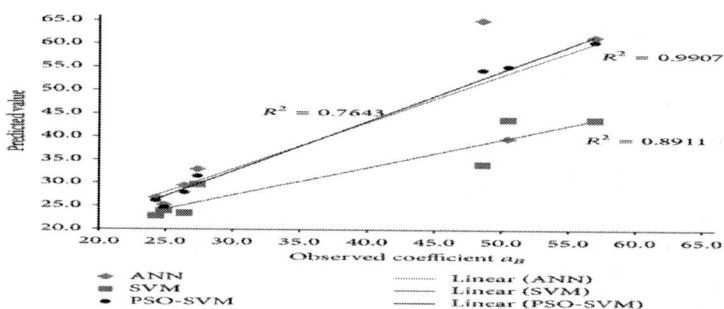

Figure 4: Predictive performance of a_B.

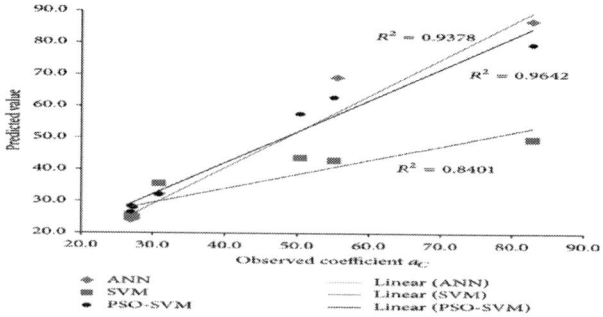

Figure 5: Predictive performance of a_C.

Predicted Error Comparison

In order to evaluate the model performance, the absolute error function and relative error function are used to compare the predictive results of the various methods used for modeling the sand rock samples. Consider

$$\text{Absolute error} = |\text{Predicted value} - \text{observed value}|,$$

$$\text{Relative error}$$

$$= \frac{|\text{Predicted value} - \text{observed value}|}{\text{Observed value}} \times 100\%.$$

(8)

The absolute prediction errors (AE) and relative prediction errors (RE) are given in Table 6, respectively. The average prediction errors of all the approaches are shown in Figure 6 for the three rock compressibility coefficients. From these results, it is interesting to see that the prediction errors of the test samples 3, 4, and 7 are much larger than those of the other test samples in all the used techniques. Generally, the prediction errors of ANN approach are the biggest and that of the PSO-SVM approach is the smallest. The average prediction error bar in Figure 6 has shown this more apparently.

Table 6: Relative prediction error of different models for the test samples

Test sample no.	ANN prediction						SVM prediction						PSO-SVM prediction					
	a_A		a_B		a_C		a_A		a_B		a_C		a_A		a_B		a_C	
	AE	RE	AE	RE	AE	RE	AE	RE	AE	RE	AE	RE	AE	RE	AE	RE	AE	RE
1	1.0	3.8	0.3	1.2	2.7	10.0	2.8	10.8	1.0	4.0	2.2	8.1	1.1	4.2	0.3	1.2	1.5	5.6
2	1.0	3.2	3.0	11.4	1.7	6.2	4.2	13.5	3.0	11.4	2.3	8.4	1.1	3.5	1.5	5.7	0.7	2.6
3	25.5	74.1	16.5	34.0	4.1	4.9	3.9	11.3	14.5	29.8	33.1	39.9	7.8	22.7	5.8	11.9	3.3	4.0
4	5.9	12.3	10.8	21.4	6.5	12.9	6.4	13.4	6.7	13.3	6.6	13.1	5.9	12.3	4.7	9.3	7.3	14.5
5	0.9	3.1	2.5	10.3	2.9	10.7	1.5	5.2	1.5	6.2	1.4	5.2	1.1	3.8	1.8	7.4	0.5	1.9
6	4.3	14.1	5.6	20.4	1.4	4.5	2.8	9.2	2.2	8.0	4.7	15.2	0.7	2.3	4.1	15.0	1.2	3.9
7	3.2	5.5	4.6	8.1	14.8	26.9	17.9	30.8	13.2	23.2	12.1	22.0	3.0	5.2	3.6	6.3	7.8	14.2
Ave.	5.97	16.61	6.19	15.24	4.87	10.89	5.64	13.46	6.01	13.69	8.91	15.99	2.96	7.72	3.11	8.12	3.19	6.64

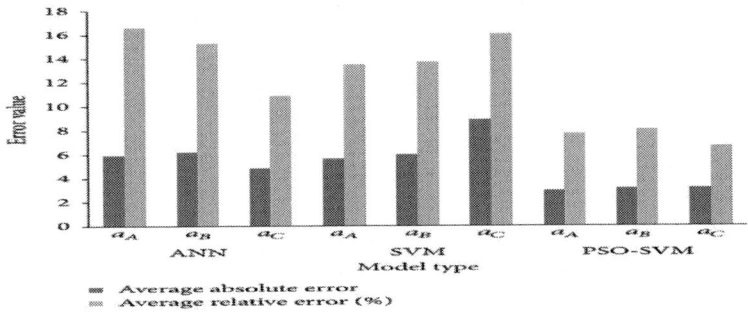

Figure 6: Average predictive errors of the models.

The prediction errors of the PSO-SVMs are nearly half of those of the ANNs. Despite this, it does not necessarily mean that the ANNs are not suitable for estimating the rock compressibility coefficients. In the modeling of ANNs approach, only a pair of initialized parameter values of the networks is given and no optimization techniques are applied to find optical parameters. In the simple SVM modeling, again, no optimizing algorithms are applied to obtain the optical penalty parameter and kernel parameters. While in the PSO-SVM modeling, the parameters in the SVM model are optimized by PSO algorithm and then a very good predictive performance is obtained. Therefore, optimization of the parameters in SVM is critical to give a good predictive performance. And the PSO-SVM is proven to perform much better than the ANN and simple SVM in the estimation of ELC of sandstone.

As mention above, doing such fundamental experiments will cost too much and sometimes even quite difficult. The introduced methods for determining the elastic parameters of porous rock materials can in some sense be appropriate to obtain such parameter as the elastic compressibility values. If a certain number of samples have been done, then it is only needed to measure some related physical features and parameters that are easily tested. The elastic compressibility can be estimated with good accuracy using the PSO-SVM method. Therefore, this technique is feasible and could be used as a potential tool for the estimation of elastic parameters of porous rock.

CONCLUSIONS

Based on the results obtained, conclusions can be made as follows. (1)The elastic compressibility of sandstone is found to have linear relations with the rock minerals, applied pressure, and pores by a linear regression analysis. Other parameters are excluded by the stepwise regression and thus can be considered not vulnerable in the estimation. (2)The predictive performances obtained by the ANN, SVM, and PSO-SVM prove that these techniques are feasible and appropriate for the estimation elastic compressibility of sandstone and can be applied to the estimation of other elastic parameters of porous rock material based on the mineral compositions and microstructural features.(3) The PSO-SVM is found to have the best predictive performance among the applied models in the estimation. It can be used as an alternative potential tool for evaluation of many other parameters of rock materials.

Nevertheless, this approach is developed and implemented based on the collected data samples. The more the samples collected are the more accurate results this technique will produce. In future, this approach is to be validated by more data samples with variability features to show its generalization ability in the estimation of elastic parameters of porous rocks.

ACKNOWLEDGMENTS

The present work was jointly supported by China Key Basic Research Program (973 Program, no. 2011CB013504) and China Natural Science Foundation (no. 11272114).

REFERENCES

1.   R. Yoshinaka, M. Osada, H. Park, T. Sasaki, and K. Sasaki, "Practical determination of mechanical design parameters of intact rock considering scale effect," Engineering Geology, vol. 96, no. 3-4, pp. 173–186, 2008. · ·

2.   W. Lin, Mechanical Properties of Mesaverde Sandstone and Shale at High Pressures, Lawrence livermore national laboratory, University of California, 1983.

3.   L. Caruso, G. Simmons, and R. Wilkens, "The physical properties of a set of sandstones—part I. The samples," International Journal of Rock Mechanics and Mining Science & Geomechanics Abstracts, vol. 22, no. 6, pp. 381–392, 1985.

4.   A. Gens, A. Ledesma, and E. E. Alonso, "Estimation of parameters in geotechnical backanalysis—II. Application to a tunnel excavation problem," Computers and Geotechnics, vol. 18, no. 1, pp. 29–46, 1996. ··

5.   F. G. Bell, "The physical and mechanical properties of the fell sandstones, Northumberland, England,"Engineering Geology, vol. 12, pp. 1–29, 1978.

6.   G. N. S. Koukis and S. Papanakli, "Laboratory testing properties of sandstones," in Proceedings of the 11th Internationa Congress of the Geological Society of Greece, pp. 1695–1699, Bulletin of the geological society of Greece, Athens, Ga, USA, 2007.

7.   M. E. Mostafa, F. Soliman, H. A. Ashry, and H. Helai, "Phyfical, electrical and mechanical properties of grantitic and limestone rocks with emphasis to their chemical compnstions," Communicatlons de la Faculté des Sciences de l·Université d·Ankara. Série C, vol. 12, pp. 1–10, 1994.

8.   P. Turgut, M. I. Yesilnacar, and H. Bulut, "Physico-thermal and mechanical properties of Sanliurfa limestone, Turkey," Bulletin of Engineering Geology and the Environment, vol. 67, no. 4, pp. 485–490, 2008. ··

9.   Y. Tan, D. Huang, and Z. Zhang, "Rock mechanical property influenced by inhomogeneity," Advances in Materials Science and Engineering, vol. 2012, Article ID 418729, 9 pages, 2012. ·

10. K. Zorlu and K. E. Kasapoglu, "Determination of geomechanical properties and collapse potential of a caliche by in situ and laboratory tests," Environmental Geology, vol. 56, no. 7, pp. 1449–1459, 2009. ··

11. Y. Zhang, J. He, Y. Wei, and D. Nie, "Prediction research of deformation modulus of weak rock zone under in-situ conditions," Journal of Mountain Science, vol. 8, no. 2, pp. 345–353, 2011. ··

12. R. A. Bearman, C. A. Briggs, and T. Kojovic, "The application of rock mechanics parameters to the prediction of comminution

behaviour," Minerals Engineering, vol. 10, no. 3, pp. 255–264, 1997.

13. S. Kahraman, "Evaluation of simple methods for assessing the uniaxial compressive strength of rock,"International Journal of Rock Mechanics and Mining Sciences, vol. 38, no. 7, pp. 981–994, 2001. ··

14. C. Gokceoglu, H. Sonmez, and A. Kayabasi, "Predicting the deformation moduli of rock masses,"International Journal of Rock Mechanics and Mining Sciences, vol. 40, no. 5, pp. 701–710, 2003. ··

15. A. Kayabasi, C. Gokceoglu, and M. Ercanoglu, "Estimating the deformation modulus of rock masses: a comparative study," International Journal of Rock Mechanics and Mining Sciences, vol. 40, no. 1, pp. 55–63, 2003. ··

16. M. Calvello and R. J. Finno, "Selecting parameters to optimize in model calibration by inverse analysis,"Computers and Geotechnics, vol. 31, no. 5, pp. 411–425, 2004.

17. B.-S. Chun, W. R. Ryu, M. Sagong, and J.-N. Do, "Indirect estimation of the rock deformation modulus based on polynomial and multiple regression analyses of the RMR system," International Journal of Rock Mechanics and Mining Sciences, vol. 46, no. 3, pp. 649–658, 2009. ··

18. N. Babanouri, H. Mansouri, S. K. Nasab, and M. Bahaadini, "A coupled method to study blast wave propagation in fractured rock masses and estimate unknown properties," Computers and Geotechnics, vol. 49, pp. 134–142, 2013. ·

19. X. Feng, Introduction of Intelligent Rock Mechanics, Science Press, Beijing, China, 2000.

20. Q. Zhang and J. Song, "Predicting mechanical behaviors of rock or rock engineering by neural network,"Chinese Journal of Rock Mechanics and Engineering, vol. 11, no. 1, pp. 35–43, 1992.

21. B.-R. Chen, X.-T. Feng, X.-L. Ding, and P. Xu, "Back analysis on rheological parameters based on pattern-genetic-neural network," Chinese Journal of Rock Mechanics and Engineering, vol. 24, no. 4, pp. 553–558, 2005.

22. W.-Y. Xu, F. Xu, and D.-W. Liu, "Study and application of displacement time series forecast based on APSO-WLSSVM,"

Chinese Journal of Geotechnical Engineering, vol. 31, no. 3, pp. 313–318, 2009.

23. Z.-B. Liu, W.-Y. Xu, F. Xu, and L.-W. Wang, "Mechanical parameter analysis of rock material via PLSR model," Key Engineering Materials, vol. 467–469, pp. 1826–1831, 2011. ··

24. Z. Liu, W. Xu, and J. Shao, "Gaussian process based approach for application on landslide displacement analysis and prediction," Computer Modeling in Engineering & Sciences, vol. 84, no. 2, pp. 99–122, 2012.

25. Z. Liu, J. Shao, W. Xu, and Y. Meng, "Prediction of rock burst classification using the technique of cloud models with attribution weight," Natural Hazards, vol. 68, no. 2, pp. 549–568, 2013.

26. C. Cortes and V. Vapnik, "Support-vector networks," Machine Learning, vol. 20, no. 3, pp. 273–297, 1995. ··

27. R. Gholami, A. R. Shahraki, and M. Jamali Paghaleh, "Prediction of hydrocarbon reservoirs permeability using support vector machine," Mathematical Problems in Engineering, vol. 2012, Article ID 670723, 18 pages, 2012. ··

28. S. S. Keerthi, V. Sindhwani, and O. Chapelle, "An efficient method for gradient-based adaptation of hyperparameters in svm models," in Advances in Neural Information Processing Systems Cambridge, B. Schölkopf, J. Platt, and T. Hoffman, Eds., pp. 673–680, MIT Press, 2007.

29. H. N. Hall, "Compressibility of reservoir rocks," Journal of Petroleum Technology, vol. 5, no. 1, pp. 17–19, 1953.

30. G. H. Newman, "Pore-volume compressibility of consolidated, friable, and unconsolidated reservoir rocks under hydrostatic loading," Journal of Petroleum Technology, vol. 25, pp. 129–134, 1973.

31. X. N. Xue and M. Li, "New calculating method of rock compressibility coeffcients petroleum geology and oilfield," Development in Daqing, vol. 30, no. 3, pp. 110–112, 2011.

32. T. M. Wissler and G. Simmons, "The physical properties of a set of sandstones-Part II. Permanent and elastic strains during hydrostatic compression to 200 MPa," International Journal of Rock Mechanics and Mining Sciences and, vol. 22, no. 6, pp. 393–406, 1985.

33. R. W. Zimmerman, W. H. Somerton, and M. S. King, "Compressibility of porous rocks," Journal of Geophysical Research, vol. 91, no. 12, pp. 765–712, 1986.

34. C. Li, X. Chen, and Z. Du, "A new relationship of rock compressibility with porosity," in Proceedings of the SPE Asia Pacific Oil and Gas Conference and Exhibition (APOGCE ‹04), pp. 163–167, October 2004.

35. A. A. Jalalh, "Compressibility of porous rocks: part II. New relationships," Acta Geophysica, vol. 54, no. 4, pp. 399–412, 2006. ··

36. A. Suman, Uncertainties in Rock Pore Compressibility and Sersmic History Matching, Stanford Stanford University, 2009.

37. J. Kennedy and R. Eberhart, "Particle swarm optimization," in Proceedings of the IEEE International Conference on Neural Networks, pp. 1942–1948, December 1995.

38. R. B. Darlington, Regression and Linear Models, Mcgraw-Hill College, New York, NY, USA, 1990.

39. S. Awasthi, General Stepwise Regression (GSR), University of Texas at Arlington.

Chapter 2

The Effect of Pore Volume of Hard Coals on Their Susceptibility to Spontaneous Combustion

Agnieszka Dudzi ska

Mines Ventilation Department, Central Mining Institute, Plac Gwarków, 40-166 Katowice, Poland

ABSTRACT

In this paper the results of the experimental studies on a relationship between pore volume of hard coals and their tendency to spontaneous combustion are presented. Pore volumes were determined by the gas adsorption method and spontaneous combustion tendencies of coals were evaluated by determination of the spontaneous combustion indexes Sz^a and Sz^a on the basis of the current Polish standards. An increase in the spontaneous combustion susceptibility of coal occurs in the case of the rise both in micropore volumes and in macropore surfaces. Porosity of coal strongly affects the possibility of oxygen diffusion into the micropores of coal located inside its porous structure. The volume of coal micropores determined on the basis of the carbon dioxide adsorption isotherms can serve as an indicator of a susceptibility of coal to spontaneous combustion.

INTRODUCTION

Hard coal is considered to be an organic polymer with a well-developed system of pores [1]. In this system one can distinguish submicro- and micropores and mesopores, as well as macropores. In a number of published research studies it has been suggested that micro- and submicropores are located mainly in the aromatic polymer and constitute the primary absorptive part of hard coal [2]. It appears that macropores are found on the edge of the polymer where they are formed as a result of the alicyclic and aliphatic hydrocarbons combination [2]. In the diffusion processes macropores are treated as transport arteries for the molecules of gas penetrating through mesopores into the micro- and submicropore system [3].

Self-heating of coal present in the deposit or pile can occur when the adequate amount of atmospheric oxygen reaches the sorptive system of the coal structure and stays in contact with the organic components. The process of self-heating of coal is influenced by many factors, including deposit temperature, the degree of coal fragmentation, type of coal and its rank, and moisture and ash content of coal, as well as maceral composition. Lower-rank coals having on their surface the large amounts of reactive groups containing oxygen adsorb more oxygen in the process of oxidation compared with coals of higher rank characterized by more organized structure and lower porosity [4]. The structure of hard coal is one of the factors that determines a susceptibility of coal to spontaneous combustion and influences the process of the self-heating of coal. Probably, energetic centers on the coal surface and oxygen-hydrogen groups are of great importance in this process. During the process of coal oxidation in the first place the following formations are oxidized: $-CH_{al}$, $-CH_{ar}$, $=CH_2$, $-CH_3$, and then finally multicore condensed aromatic systems [3]. Oxygen consumption will also be determined by its transport to the inside of the microporous structure of coals. Hence, of particular significance will also be a system of transport pores consisting of meso- and macropores. According to some authors, it is exactly meso- and macropores that play a great role in the processes of coal oxidation [5]. Although micropores comprise over 90% of the total pore volume, not always their large volume affects an increase in the oxidation rate of coal [6]. The kinetics of the reaction of coal oxidation depends mainly on the development of the solid-phase surface, the presence of active centers on it, and the presence and size of pores enabling transport of the gas phase. The structure of coal determines its accessibility for gases, including atmospheric oxygen, and this is one of the conditions of spontaneous ignition. Coal oxidation takes place both on the outside surfaces of the grains and the inside of the porous structure [7].

In this study, the authors analyzed the influence of the hard coals structure on their tendency to spontaneous combustion, using the parameter measured by gas adsorption method volume of pores. Adsorption of vapours and gases conducted on hard coals is a source of much information concerning mainly surface and porosity of coals, which are used to describe their properties. Many papers have been published in the areas of gas adsorption such as carbon dioxide, methane, water, carbon monoxide, hydrogen, and hydrocarbons [8–15].

CURRENTLY USED METHODS FOR DETERMINATION OF A SUSCEPTIBILITY OF COAL TO SPONTANEOUS COMBUSTION

The self-heating phenomenon occurs when the heat emitted as a result of contact between hard coal and oxygen is accumulated. In consequence of this accumulation heating up of coal in the deposit takes place, and then, under favorable conditions, the temperature of coal rises and after reaching the ignition temperature an endogenous fire can occur. Endogenous fires are a real threat both to safety of mining crews and to individual processes regarding coal, as from mining, through transport, to recycling; they also cause huge financial losses. The phenomenon of self-heating of hard coals was studied repeatedly, and yet the course of this complicated process is still not entirely known. The complexity of the process of coal oxidation in the deposit results from the impact of many factors associated with this process, whose role in it is not always easy to describe or interpret. From the work safety point of view the proper evaluation of the process of self-heating of coal is extremely crucial as it enables full control over the course of this phenomenon and thus, if needed, an inhibition of its further development in order to prevent dangerous and costly fires. One of the elements of fire hazard assessment is to study the susceptibility of coals to spontaneous combustion. For many years attempts have been made to develop methods for determining the self-ignition tendencies of hard coals. A number of hypotheses and theories of this process have been put forward, including the pyrite theory, phenol theory, and coal-oxygen complex theory. There have been distinguished four groups of methods based on adiabatic calorimetry, isothermal calorimetry, sorption of oxygen, and the rate of temperature rise of a sample in relation to the temperature of a specified reference system [16].

Among many presented methods for evaluation of the spontaneous combustion tendency of coal much space has been devoted to the method of adiabatic calorimetry [17, 18], which involves measuring the amount of heat emitted under certain fixed external conditions and recording changes in temperature of a coal sample or in the amount of evaporating moisture. This method is useful for long-term processes associated with low-temperature oxidation of coals. In the method of oxygen sorption, in turn, the amount of oxygen used in the process of coal oxidation during a laboratory test is the parameter which determines the qualification of a given coal for a group of low, medium, or high spontaneous combustion susceptibility [19]. In other methods, the critical temperature, which is the lowest temperature at which under laboratory conditions self-heating of coal takes place, is determined and then this temperature is a criterion for the classification of coals as more or less prone to spontaneous combustion [20].

In Poland the currently used method for examination of hard coals tendencies to spontaneous combustion is the Olpinski method covered by the standard PN-93/G-04558.

EXPERIMENTAL PART

Laboratory tests were carried out on 18 samples of hard coals having varied degrees of metamorphism and different susceptibility to spontaneous combustion, collected from exploitable coal seams in Polish mines. From a laboratory sample of each coal, prepared in accordance with the standard PN-90/G-04502, by means of Fritsch sieves a grain class of 0,063–0,075 mm was isolated. Coal samples of this grain size were examined to determine the values of coal spontaneous combustion indexes and tested for adsorption.

Determination of the Spontaneous Combustion Index of Coal

Determination of the spontaneous combustion index of coal is performed on the basis of the standard PN-93/G-04558 "Hard Coal Determination of the spontaneous combustion index." The principle of this method lies in the continuous measurement of the temperature of the pellet made from tested coal, introduced into the air stream at 546 K and 463 K and in the determination of the rate of temperature increase of this pellet at adiabatic points. Coal pellets are made from a specially isolated grain class of 0.063–0.075 mm. In this method spontaneous combustion index Sz^a [K/min] is determined—that is, the rate of temperature increase of coal after its oxidation in air at 546 K in an adiabatic (reaction) chamber and similarly index Sz^a [K/min] at 463 K. The activation energy is also determined—that is, the minimum energy that a group of molecules must have in order for a reaction to occur, expressed by a value of A in the following Arrhenius equation:

$$k = k_0 e^{-A/RT},$$

(1)

describing the relationship between a reaction rate constant (k) and temperature (T), where A is the activation energy, R is the universal gas constant e—2,718, and k_0 is preexponential factor.

The activation energy is expressed by the following formula:

$$A = 96{,}79 \log \frac{Sz^a}{Sz^{a'}}, \text{kJ/mol.}$$

(2)

On the basis of the values of coal spontaneous combustion index Sz^a and the activation energy a division into five groups of spontaneous combustion according to the standard PN-93/G-04558 has been made. This division is shown in Table 1.

Table 1: Classification of coal by a susceptibility of spontaneous combustion

Spontaneous combustion index Sz^a (K/min)	Activation energy of coal oxidation A (kJ/mol)	Group of spontaneous combustion	Evaluation of a susceptibility of coal to spontaneous combustion
≤80	>67	I	Coal of a very low spontaneous combustion susceptibility
	46–67	II	Coal of a low spontaneous combustion susceptibility
	<46	III	Coal of a medium spontaneous combustion susceptibility
>80 and ≤100	>42		
	42 or less	IV	Coal of a high spontaneous combustion susceptibility
>100 and ≤120	>34		
	34 or less	V	Coal of a very high spontaneous combustion susceptibility
>120	No standarization		

The values of spontaneous combustion indexes and the activation energy for individual coal samples with the corresponding group of spontaneous combustion are shown in Table 2.

Table 2: The values of spontaneous combustion indexes and the activation energy for individual coal samples with the corresponding group of spontaneous combustion

Samples	Spontaneous combustion index Sz^a (K/min)^a	Spontaneous combustion index Sz$^{a'}$ (K/min)	Activation energy of coal oxidation (kJ/mol)	Group of spontaneous combustion
Rydułtowy s. 703/1	12	4	46	II
Marcel s. 712/1-2	13	3	62	II
Borynia s. 415	46	8	73	I
Chwałowice s. 405	46	7	79	I
Zofiówka s. 407	55	11	68	I
Marcel s. 507	57	12	65	II
Bielszowice s. 405/2	71	17	60	II
Jankowice s. 405	84	32	40	IV
Śląsk s. 510	91	39	36	IV
Sośnica s. 413	92	31	46	III
Szczygłowice s. 405	92	32	44	III
Polska-Wirek s. 510	91	40	35	IV
Staszic s. 402	94	34	43	III
Wesoła s. 501B	102	32	49	IV
Piast s. 209	103	46	34	V
Mysłowice s. 501	104	45	35	IV
Piast s. 206	105	48	33	V
Sobieski s. 209	150	63	36	V

Adsorption Tests

To determine the pore volume of coal the adsorption tests on examined coal samples were performed, using carbon dioxide and nitrogen as adsorbates. The measurements of the adsorption isotherms of carbon dioxide were conducted at 298 K and of the nitrogen adsorption isotherms at 77.5 K. The adsorption tests were performed applying the volumetric method with the use of an apparatus ASAP 2010 Accelerated Surface Area and Porosimetry Analyzer by Micromeritics in the range of pressure 0-0.1 MPa. Prior to measurements hard coal samples crushed to a grain class of 0.063–0.075 mm were degassed under a high vacuum of 10^{-7} Pa. During degassing they were flushed with helium for easier removal of gas residuals contaminating the coal surface.

The results obtained from the adsorption measurements comprising micropore volumes calculated according to the Dubinin-Radushkevich model based on the carbon dioxide adsorption isotherms and macropore surfaces area calculated according to the BET model based on the nitrogen adsorption isotherms are shown in Table 3. Those results were next used to plot values of spontaneous combustion indexes Sz^a and Sz^a both as a function of surface area of macropores and as a function of volume of micropores present in coal.

Table 3: The values of micropores volume and surface area of macropores for individual coal samples

Samples	Micropores volume, cm³/g	Surface area of macropores, m²/g
Rydułtowy s. 703/1	0.033	2.53
Marcel s. 712/1-2	0.043	2.07
Borynia s. 415	0.047	2.21
Chwałowice s. 405	0.048	3.00
Zofiówka s. 407	0.041	3.92
Marcel s. 507	0.038	3.84
Bielszowice s. 405/2	0.044	2.10
Jankowice s. 405	0.056	2.02
Śląsk s. 510	0.049	3.55
Sośnica s. 413	0.050	2.91
Szczygłowice s. 405	0.055	3.85

Polska-Wirek s. 510	0.049	4.88
Staszic s. 402	0.055	3.85
Wesoła s. 501B	0.059	6.03
Piast s. 209	0.051	4.79
Mysłowice s. 501	0.051	3.82
Piast s. 206	0.072	8.99
Sobieski s. 209	0.075	13.52

RESULTS AND DISCUSSION

Effect of Micropore Volume on a Susceptibility of Coals to Spontaneous Combustion

The main purpose of this part of the study is to determine the relationship between micro- and submicropore volumes of coal and the values of indexes characterizing susceptibility of coals to spontaneous combustion. Spontaneous combustion of coal is possible in the case of the reaction of specific groups in coal with oxygen. The number of coal active centers depends on the development of the inside micropores surface area and on the accessibility of active centers located in the reactive part of coal for oxygen. Micropore volume becomes then an important factor determining the ability of coal to absorb atmospheric oxygen. Larger micropore volumes cause greater exposition of coal to gases, and thus it is more likely that the process of spontaneous combustion will occur.

The method of gas adsorption enables determination of micropore volumes of materials tested and the gas most commonly used for determination of hard coals micropore volumes is carbon dioxide. Its low chemical reactivity, a proper size of the molecule determined by the so-called kinetic diameter (for carbon dioxide it amounts to 0.33 nm) [21], and the low value of the activation energy are the factors favoring the possibility of carbon dioxide penetration into the pores, which are not available for other gases. According to [22, 23] carbon dioxide is the gas best adsorbed inside the microporous structure of hard coals. Not without significance is also the temperature at which the measurements of adsorption were performed −298 K, as it is the closest to the actual temperature at the bottom of the mine and thus provides an easy access of the carbon dioxide molecules to the coal structure. The adsorption of carbon dioxide takes place in the entire coal structure, both on the outer surfaces and inside the microporous structure, as confirmed in many studies [24]. Considering the CO_2 adsorption in micropores of hard coals, hydroxyl groups present in the coal structure may be of great importance, as they can electrostatically interact with the CO_2 molecule having a double dipole character.

Figure 1 shows the dependence of the values of spontaneous combustion index Sz[a] on micropore volumes, determined from the course of the CO_2 adsorption isotherms. Analyzing this relationship, it can be seen that the values of the index rise with the increasing micropore volume. This relationship is not linear but in a band form, with a rather large scatter of data. Similarly, Figure 2 presents the dependence of the values of spontaneous combustion index Sz[a] on micropore volumes. In this case data are also scattered but the values of index are lower. The index Sz[a] is determined in the process of its oxidation in air at lower temperature of 463 K. The data presented in Figures 1 and 2 distinctly show the impact of micropore volume on the values of coal spontaneous combustion indexes Sz[a] and Sz[a] symbols and thereby confirm the clear influence of the inside microporous structure of coals on their susceptibility to spontaneous combustion.

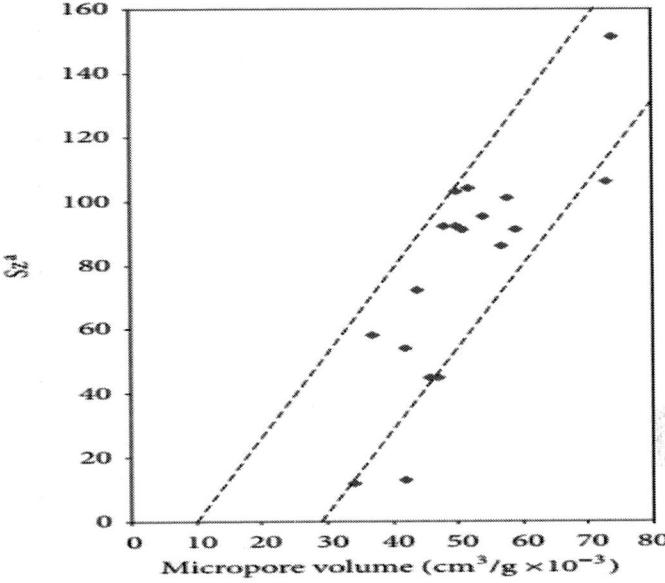

Figure 1: Value of spontaneous combustion index Sz^a of coal as a function of micropore volume.

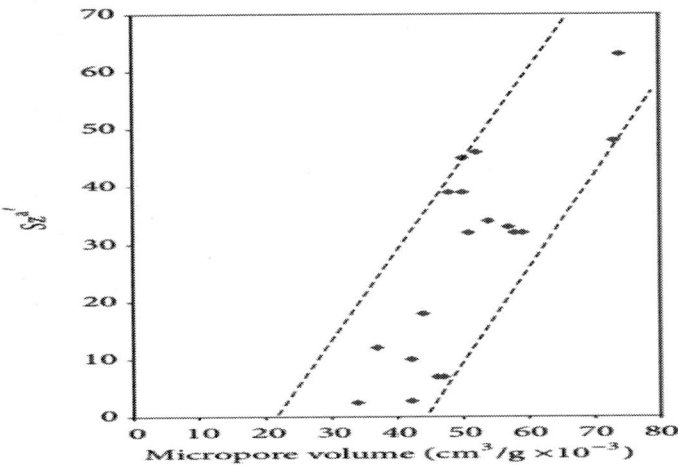

Figure 2: Value of spontaneous combustion index Sz^a of coal as a function of micropore volume.

Coals selected for tests are characterized by different values of spontaneous combustion indexes, from the lowest values characteristic of I group of spontaneous combustion up to high values of indexes corresponding to coals demonstrating high tendency to self-ignition—V group of spontaneous combustion (Table 2). The participation of micropore volume in the oxidation processes applies to all coals, from the highest-rank coals with organized structure whose susceptibility to spontaneous combustion is usually low to those with the lowest percentage of carbon and the highest tendency to spontaneous combustion. Due to a diversity of the micropore volume values obtained from CO_2 adsorption, they can be used to determine the degree of spontaneous combustion tendency of coal.

Relationship between Macropore Surface and a Susceptibility of Hard Coals to Spontaneous Combustion

Larger mesopores and macropores, the so-called transport pores, are responsible for the transport of sorbed gas (oxygen) into the deep inside of the coal structure. At the initial stage, self-heating of coal is determined by the amount of oxygen adsorbed in macropores and mesopores of the given hard coal. Some authors claim that it is macropores, treated as transport arteries for the molecules of penetrating gas, that make a significant contribution to the processes of coal oxidation [5]. According to [25], pores of a size less than 100 Å do not contribute to the reaction of low-temperature oxidation due to the lack of microdiffusion. For macropore surface area examination the measurements of nitrogen adsorption have been proposed. In many research studies on the adsorption of nitrogen conducted at 77.5 K on hard coals it has been shown that in those conditions nitrogen is adsorbed only on the inner walls of macropores. At the temperature of liquid nitrogen coal contracts what causes the narrowing of the fine capillaries, preventing the movement of the nitrogen molecules into the inside of further micropores [23]. The influence of the diffusion factor is also being taken into account. Low temperature causes a reduction in the kinetic energy of the nitrogen molecules penetrating the capillaries which stops them in narrow capillaries of the porous coal structure. In the theory of diffusion processes it is assumed that gases penetrating into micropores are first collected in macropores and then transmitted through the mesopore system into micro- and submicropores [26]. In the case of low volume of macropores not all micropores receive enough oxygen which results in slowing down the process of coal oxidation. In literature the issue concerning coal susceptibility to spontaneous combustion resulting from the value of the macropores surface area has been rarely discussed. The obtained tests results were used to plot graphs shown in Figures 3 and 4.

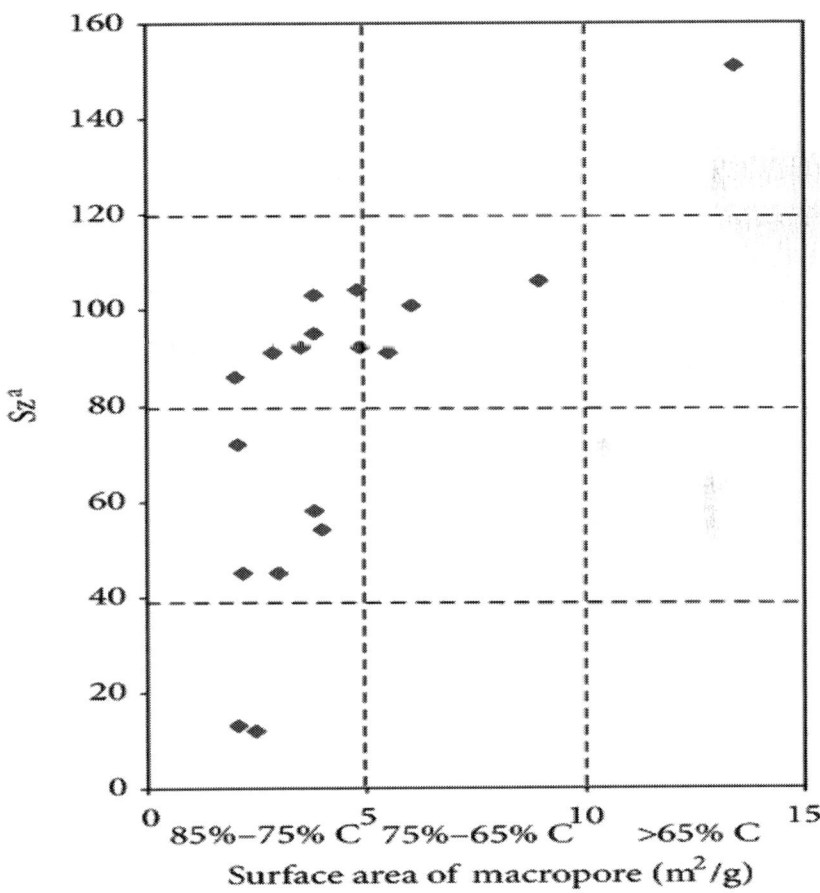

Figure 3: Value of spontaneous combustion index Sz^a of coal as a function of macropore surface area.

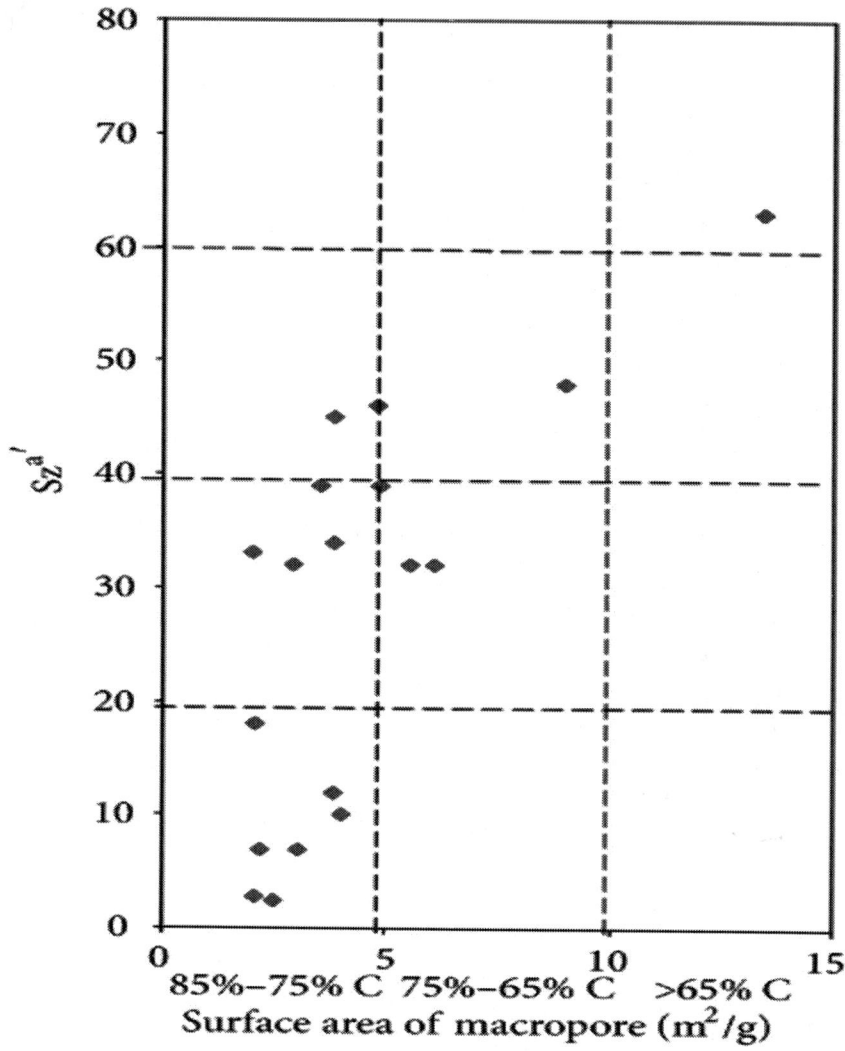

Figure 4: Value of spontaneous combustion index Sz^a of coal as a function of macropore surface area.

Figure 3 shows the dependence of the values of Sz^a index on the macropore surface area. The plotted curve features a band character. After dividing the graph into several rectangular areas a few important remarks can be made. The values of Sz^a index falling within a range of macropore surface area values of 1–5 m^2/g include most of the tested coals and show a gradual increase in the value of this index with a little change of the surface area value. A significant share of these coals demonstrates a similar macropore surface area in the order of 2-3 m^2/g, although the values of spontaneous combustion indexes of those coals are varied and comprise I–III groups of spontaneous combustion. Coals falling into this range are coals of higher rank with the carbon content of 75–85%. Probably the low value of macropore surface area can hinder the access of adsorbed gases to the inside microporous structure of coals, which can slow down the process of oxidation. It is slightly different in the case of lower-rank coals with the loose structure, whose carbon content is less than 75% C. For those coals the values of macropore surface area differ more. In this case an increase in the spontaneous combustion index is concurrent with a clear increase in the macropore surface area.

In Figure 4 an analogous graph concerning the relationship between the values of Sz^a index and macropore surface area is shown. The values of the index discussed in the range of the low values of macropore surface area (1–5 m^2/g) are distinctly different, as in the case of Sz^a index. With an increase in the surface area of macropores of more than 5 m^2/g the coals already demonstrate quite a high tendency to spontaneous combustion which, undoubtedly, is also associated with a variety of energetic centers on their surface and with a significant increase in the number of reactive groups containing oxygen.

The low values of macropore surface area are the result of physical adsorption characterized by weak dispersion force interactions occurring between the nitrogen molecules and the coal substance. The values of macropore surface area can also be used to evaluate the tendency of coal to spontaneous combustion, although the index expressed by macropore surface area seems to be better only for coals having lower carbon content (below 75% C). For coals of the higher degree of metamorphism, the values of specific surface area are similar, although the index values of those coals are very much different.

CONCLUSIONS

Summing up the presented tests results it can be concluded that there certainly is a relationship between pore volume of hard coals and their tendency to spontaneous combustion.

- Micropore volumes of coals determined on the basis of the carbon dioxide adsorption isotherms increase with the values of spontaneous combustion indexes Sz^a and Sz^a of coal according to the standard PN-93/G-04558. Coals with higher values of micropore volume are more prone to spontaneous combustion.

- Macropore surface areas determined on the basis of the nitrogen adsorption isotherms also correlate with the values of spontaneous combustion indexes. High-rank coals demonstrate a small change in the surface area of macropores in relation to changes in the values of spontaneous combustion indexes. In the case of coals of lower rank there is a clear rise in spontaneous combustion indexes with the increasing macropore surface area.

- Pore volume determined from the adsorption of carbon dioxide seems to be a better parameter that can be used to assess the susceptibility of coal to spontaneous combustion. The molecules of carbon dioxide are well adsorbed in the pore structure of coal; the temperature during measurements is similar to the actual conditions in situ and at the same time there is a greater diversity of micropore volumes compared to the macropore surface area.

REFERENCES

1. G. Ceglarska-Stefańska, A. Czapliński, P. Fudalej, and S. Hołda, "Polimeryczny model węgla kamiennego w świetle wysokociśnieniowych badań sorpcyjnych i dylatometrycznych," Archives of Mining Sciences, vol. 20, pp. 299–305, 1975.

2. C. R. Clarkson and R. M. Bustin, "Effect of pore structure and gas pressure upon the transport properties of coal: a laboratory and modeling study. 1. Isotherms and pore volume distributions," Fuel, vol. 78, no. 11, pp. 1333–1344, 1999. · ·

3. A. Czapliński, Węgiel Kamienny, Wydawnictwa AGH, Kraków, Poland, 1994.

4. Q. I. Xuyao, "Characteristics of oxygen consumption of coal at programmed temperatures," Mining Sciences and Technology, vol. 20, pp. 0372–0377, 2010.

5. Y. S. Nugroho, A. C. McIntosh, and B. M. Gibbs, "Low-temperature oxidation of single and blended coals," Fuel, vol. 79, no. 15, pp. 1951–1961, 2000. · ·

6. X. Wang, R. He, and Y. Chen, "Evolution of porous fractal properties during coal devolatilization," Fuel, vol. 87, no. 6, pp. 878–884, 2008. · ·

7. H. Wang, B. Z. Dlugogorski, and E. M. Kennedy, "Coal oxidation at low temperatures: oxygen consumption, oxidation products, reaction mechanism and kinetic modelling," Progress in Energy and Combustion Science, vol. 29, no. 6, pp. 487–513, 2003. · ·

8. O. P. Mahajan, "CO_2 surface area of coals: the 25-year paradox," Carbon, vol. 29, no. 6, pp. 735–742, 1991.

9. J. H. Levy, S. J. Day, and J. S. Killingley, "Methane capacities of Bowen Basin coals related to coal properties," Fuel, vol. 76, no. 9, pp. 813–819, 1997.

10. J. Cygankiewicz, A. Dudzińska, and M. Żyła, "Sorption and desorption of carbon monoxide in several samples of polish hard coal," Archives of Mining Sciences, vol. 52, no. 4, pp. 573–585, 2007.

11. J. Cygankiewicz, A. Dudzińska, and M. Żyła, "The relation between the size of bituminous coal particles and the sorption of carbon monoxide," Gospodarka Surowcami Mineralnymi/ Mineral Resources Management, vol. 25, no. 1, pp. 85–100, 2009.

12. M. Żyła, A. Dudzińska, and J. Cygankiewicz, "The relation between ambient temperature and sorption of carbon monoxide on bituminous coals," Gospodarka Surowcami Mineralnymi/ Mineral Resources Management, vol. 25, no. 4, pp. 33–49, 2009.

13. J. Cygankiewicz, A. Dudzińska, and M. Żyła, "Examination of sorption and desorption of hydrogen on several samples of polish hard coals," Adsorption, vol. 18, no. 3-4, pp. 189–198, 2012.

14. M. Żyła, A. Dudzińska, and J. Cygankiewicz, "The influence of disintegration of hard coal varieties of different metamorphism grade on the amount of absorbed ethane," Archives of Mining Sciences, vol. 58, no. 2, pp. 449–463, 2013.

15. A. Dudzińska, M. Żyła, and J. Cygankiewicz, "Influence of the metamorphism grade and porosity of hard coal on sorption and desorption of propane," Archives of Mining Sciences, vol. 58, no. 3, pp. 859–871, 2013.

16. A. G. Kim, "Laboratory studies on spontaneous heating of coal—a summary of information in the literature," U.S. Bureau of Mines, no. 8756, 1977.

17. J. M. Kuchta, V. R. Rowe, and D. S. Burgess, "Spontaneous combustion susceptibility of U.S. coals,"Bureau of Mines Report of Investigation, vol. 37, no. 8474, 1980.

18. T. X. Ren and M. J. Richards, "Computerised system for the study of the spontaneous combustion of coal," Mining Engineer London, vol. 154, no. 398, pp. 121–127, 1994.

19. G. Y. Qian, "Research on the use of spectrum technique for determining the liability of coal to spontaneous combustion fushun coal reserch institute of technology," P.R. China, 1987 (Chinese).

20. K. K. Feng, R. N. Chakravorty, and T. S. Cochrane, "Spontaneous combustion-a coal mining hazard,"C/M Bulletin, vol. 66, no. 738, pp. 75–84, 1973.

21. P. Chowdhury, C. Bikkina, and S. Gumma, "Gas adsorption properties of the chromium-based metal organic framework MIL-101," Journal of Physical Chemistry C, vol. 113, no. 16, pp. 6616–6621, 2009. · ·

22. K. Zarębska, P. Baran, J. Cygankiewicz, and A. Dudzińska, "Sorption of carbon dioxide on polish coals in low and elevated pressure," Fresenius. Environmental Biulletin, vol. 21, pp. 4003–4008, 2012.

23. J. Cygankiewicz, A. Dudzińska, and M. Żyła, "The effect of particle size of comminuted bituminous coal on low-temperature sorption of nitrogen and room-temperature sorption of carbon dioxide," Przemysl Chemiczny, vol. 85, no. 11, pp. 1505–1509, 2006.

24. B. M. Krooss, F. Van Bergen, Y. Gensterblum, N. Siemons, H. J. M. Pagnier, and P. David, "High-pressure methane and carbon dioxide adsorption on dry and moisture-equilibrated

Pennsylvanian coals," International Journal of Coal Geology, vol. 51, no. 2, pp. 69–92, 2002. · ·

25.   R. Kaji, Y. Hishinuma, and Y. Nakamura, "Low temperature oxidation of coals. Effects of pore structure and coal composition," Fuel, vol. 64, no. 3, pp. 297–302, 1985.

26.   D. Prinz, W. Pyckhout-Hintzen, and R. Littke, "Development of the meso- and macroporous structure of coals with rank as analysed with small angle neutron scattering and adsorption experiments," Fuel, vol. 83, no. 4-5, pp. 547–556, 2004.

Chapter 3

Mining and Seasonal Variation of the Metals Concentration in the Puyango River Basin—Ecuador

Maria Eugenia Garcia[1], Oscar Betancourt[2], Edwin Cueva[2], and Jean Remy D. Gimaraes[3]

[1]Chemical Research Institute, San Andrés University, La Paz, Bolivia
[2]Health Environment Development Foundation (FUNSAD), Quito, Ecuador; 3Rio de Janeiro University, Rio de Janeiro, Brasil

ABSTRACT

The Puyango River Basin covers approximately an area of 4400 km^2, it is located in Southern of Ecuador, with Calera and Amarillo rivers as tributaries. In this region, one of the main activities is small scale gold and silver mining. Currently there are 110 processing plants on the bank of Calera and Amarillo rivers, causing a significant degradation of natural resources. A seasonal comparison of metal concentrations in surface water, sediments and particulate matter from the Puyango River and its effluents is made. It was done a differentiation between natural contaminations with the anthropogenic one generated by mining activity. Samples were taken during dry season (2004) and rainy season (2006), and analyzed physicochemical parameters, anions and cations and the concentrations of heavy metals. The results show a clear influence of gold mining in Puyango River contamination, starting with its tributaries, Calera and Amarillo rivers, which have the highest concentrations of heavy metals from the basin, corresponding with the location of the mineral processing plants.

INTRODUCTION

The study area is located at East between 645513.93 and 9583432.64 and 9558839.96 West between 569100.97 at Southwest of Ecuador and covers a total area of approximately 4400 km². This area is located in the Puyango River Basin with its many tributaries. There is an altitudinal gradient from 2000 to approximately 100 m.a.s.l.

The geographical area covered by the project area is located in the foothills of the Western Cordillera of the Andes to the west; this feature is perhaps one of the dominant facts of determining the sector.

The winds that bring moist air masses from the Pacific Ocean, the humid air masses that move from the Pacific penetrate the middle and upper basin of the river to the area Puyango imposing a maritime regime, determining two seasons: a rainy winter from January to April and a dry summer May to November or December, where the average annual rainfall is 1160 mm.

The main axis of the basin is the Pindo-Puyango River, wich from the mining area goes trough 120 km until the Peru border. The Puyango River basin covers different towns of Loja and El Oro provinces, where is developed activities as agriculture, livestock, mining and tourism. The most relevant for this study are Zaruma and Portovelo cities.

Mining is the most important activity in the highest zone of the Basin (Zaruma-Portovelo) and exists since pre-Columbian times. Ecuadorian historians speak that mines were known and exploited since Inca [1], this operation continued in the colonial period. Viceroy Francisco de Toledo chosen these sites in 1570 for mining exploitation in Ecuador.

This activity was unsuccessful because the Spanish conquerors chose for mining Bolivia and Peru [2], with consequent environmental impacts inherited in the colonial period [3].

Mining was reactivated in 1896 trough the entrance of the South American Development Company (SADCO), a subsidiary of a US Vanderbilt Company [4], with which appears an industrial form of exploitation mining in Ecuador. The company left the country a few years after the end of II World War (1950). As a result of multiple economic, social and political aspects of Ecuador, in 1980's in the southwest and southeast there were several forms of mining production, among which is the extraction of gold by amalgamation with mercury. At begin of 2000's claims of mining companies were consolidating, process that leaves aside the social and environmental implications [5]. In the last decade reinforces the small-scale mining with concessions of 150 hectares and extractions of approximately 100 tons per day and investments to achieve a maximum of one million dollars [5].

Various forms of mining activities were consolidated in the highest part of the Puyango River Basin (Zaruma and Portovelo cities) with individual, family and mining companies forms of productive organizations. Mining in this area is characterized by metal type (gold) as tunnel mining, located in the Cordillera Vizacaya. There are also a hundred processing plants distributed in Calera and Yellow Rivers, tributaries of Puyango.

The labor process in mining has two great moments; one is made in the mines, which gives the raw ore and the other, in processing plants [6].

Processing plants are located on the bank of rivers Calera and Amarillo and various processes are performed for enriching raw ore to obtain gold or silver. It begins with size reduction of the raw ore using chilean mills, continues with gravimetric concentration until obtain a concentrate, it enters to the phase of amalgamation and burning with mercury emissions to air and water. A great part of the useful material goes to the phase of leaching with use of cyanide. It is used pools of leaching or agitation that after extract the metals for leaching; tailings are eliminated to the rivers. The final phases of dressing techniques include calcination, smelting and refining that, like other phases, it is used many chemicals substances (zinc, nitric acid) that contaminate the environment by different ways [7].

Mining in these processing plants pollutes rivers Calera and Amarillo by throwing cyanidation tailings, foam flotation, nitric acid residues and other residues. The most worrying because their impact on human health is the presence of heavy metals like mercury, lead, zinc, arsenic, cadmium and manganese, in addition to high concentrations of cyanide [8]. In addition is the soil erosion from the farming along the Puyango River Basin.

There have been few studies about the presence of metals resulting from mining in the Puyango River Basin [9-12], nevertheless, all talked about the influence of mining in the presence of metals in different parts of the rivers. These studies where focus on a circumscribed area of the basin, especially in the upper basin, the present study was conducted throughout the basin, including the origins of the rivers, the mining area, the middle Basin and the low Basin, near to the border with the neighboring country Peru. As well as a study in a neighboring country, Colombia [13], we also sampled the river in dry and rainy season.

It is well known that the geodynamics of the metals is complex, that is why several studies in America and around the world have considered important the determination of metals, especially mercury in particulate matter and/or sediments and material dissolved in water [14-18].

In addition, the present study has the particularity that includes the principal axes and tributaries of the Puyango River Basin. This allowed to distinguish the presence of heavy metals from mining of which comes from soil erosion of the riverbanks. The influence of seasons (dry and rainy), pH and conductivity were related with contamination. The metals selected for study are the most important to human health (lead, mercury, manganese and arsenic). We also studied the relationship between concentrations of cyanide in the river with the bacterial activity and potential methylation of mercury, about which there is another article [8].

This article focuses on the analysis of seasonal variation of physicochemical parameters and heavy metal concentrations in both seasons (dry and raining seasons); in surface waters, sediments and particulate matter, and identifying the influence of mining pollution of mercury, lead, manganese and arsenic.

The results have served to take measures to prevent health alterations, for example, discussing with communities that the river has high levels

Mining and Seasonal Variation of the Metals Concentration in the...

51

of metals in particulate matter. The result of these dialogues is the use of water filters to retain the particulate material. Subsequently and as a result of the strengthening of their organizations and networks of plants obtained drinking water. They are the uses of research results to improve the quality of people life.

METHODOLOGY

We designed an environmental sampling plan along the entire Puyango River Basin. Selection of sampling points was made based on criteria of representativeness, access to 20 sampling points, sites affected by mining and sites without much anthropogenic influence. Was considered also geological issues and geography of the Puyango River Basin including all tributaries of the main axis of the river. Sampling was conducted in the main transect and all tributaries over a length of approximately 160 kilometers. Table 1 shows the sampling points with their codes, name of the river and their geographical altitudes.

Water, sediments, particulate matter and soil were taken in each point and in two seasons (dry season, 2001 and rainy season, 2006). Physicochemical parameters, major cations and anions, metals and metalloids (Pb, Hg, Mn and As) were analyzed.

Triplicate samples were taken for subsequent statisticcal analysis. At pristine points, which corresponding to highest part of the river Basin (points 1 to 4), we verified the absence of mining activities and other anthropogenic activities, basic aspect for comparison the results with those found in the other sampling points. Other points were selected at convergence of tributaries with the main axis of the river and after that union.

Table 1: Sampling points, Puyango River Basin

Code	Description	Altitude (m.a.s.l.)
P1	Luis River (point of reference)	1821
P2	Ambocas River (point of reference)	970
P3	Amarillo River (point of reference)	1578

P4	Calera River (point of reference)	1553
P5	Amarillo River after the union of Calera	571
P6	Pindo River after Amarillo River	593
P7	Pindo River close Hydrologic Station	529
P8	Piñas River	605
P9	Pindo River before Yaguachi	472
P10	Yaguachi River	462
P11	Moromoro River	448
P12	Puyango River before Balsas River	388
P13	Balsas River	380
P14	Puyango River before Marcabelí River	345
P15	Marcabelí River	350
P16	Puyango River before Q. Tunima	333
P17	Tunima Ravine	337
P18	Cochurco Ravine	370
P19	Puyango River (Puyango Viejo)	300
P20	Puyango River (Gramadal-Las Vegas)	155

As Figure 1 shows, the selected sampling points allow to have a better idea of the dynamics of contamination or dilution of each Puyango River Basin tributaries. The sampling points located immediately after the processing plants were taken as a special references for analysis. Overall, the sampling network consists of 20 points, including PindoPuyango River and tributaries of the middle and lower Basin.

Montgomery [19] protocols, enriched by Roulet [20] were followed for sampling, which in summary is: Sampling of surface water with a peristaltic manual pump, operated by battery, with hoses treated previously with nitric acid and rinsed with water "milliQ". Water passed through a filter of 0.2 μm EPM2000, samples were standardized and collected in bottles of different materials (dark glass, polypropylene and teflon), depending of type of analysis. Bottles also were treated with nitric acid and with ultrapure water. Immediately after collecting, the samples were added ultra pure nitric acid and stored under refrigeration.

Polypropylene flasks of 1 liter were used for particulate material analysis, collecting water from the river directly, then, the material was captured in glass fiber filters, previously heated to 300°C, cooled and heavy (0.7 µm, GFF Whatman, Brentford, London, United Kingdom). It was used also polycarbonate membrane of 0.45 µm. Sediments were collected in representative points, using Petri boxes of 10 cm diameter and then were passed through a nylon sieve of 250 µm. Both, the sediments and particulate matter were dried in an oven at 45°C [21].

Conductivity, temperature and pH were measured in field with Oakton PC 300, after respective calibration; Eh was also measured in field with CONSORT equipment.

Lead, arsenic and manganese were analyzed by atomic absorption spectrophotometry with graphite furnace (GFAAS), AAnalyst 100. Mercury in water was analyzed by spectrophotometry and cold vapor atomic fluorescence (CVAFS) [22].

For analyzes of heavy metals in sediments and particulate material, samples were dried at 45°C and digested with a mixture of hydrochloric acid and nitric acid and finally they were analyzed through GFAAS.

The detection levels for solid materials (sediment) were of 5.5 µg/kg for lead, 4 µg/kg for arsenic and 5 µg/kg for manganese. Were taken as a reference solution SPS-WW1, Batch 108.

RESULTS

Seasonal variation was observed in both physicochemical parameters and heavy metal concentrations in surface water, sediments and particulate matter. There is a clear influence of mining activities in these concentrations; following paragraphs show the details

PH AND CONDUCTIVITY

Surface waters: values of pH are higher in dry season than in rainy season, due to a neutralization effect by the increase of precipitations, however, in some cases pH values are increased because the lime (CaO), used to precipitate metals and to avoid toxic effects of cyanide, used in gold leaching processes.

In Amarillo River, after the union with Calera River (point 5) and Cochurco Ravine (Point 8), pH is alkaline (8.92) due to higher concentrations of elements in dry season, the presence of carbonates and bicarbonates and probably because the presence of calcium oxide, both groups coming from tailings thrown into the rivers. It is important to remember that Point 5 is the area immediately located after the biggest zone of the processing plants. Puyango River receives the largest discharge of tailings and effluents from these plants.

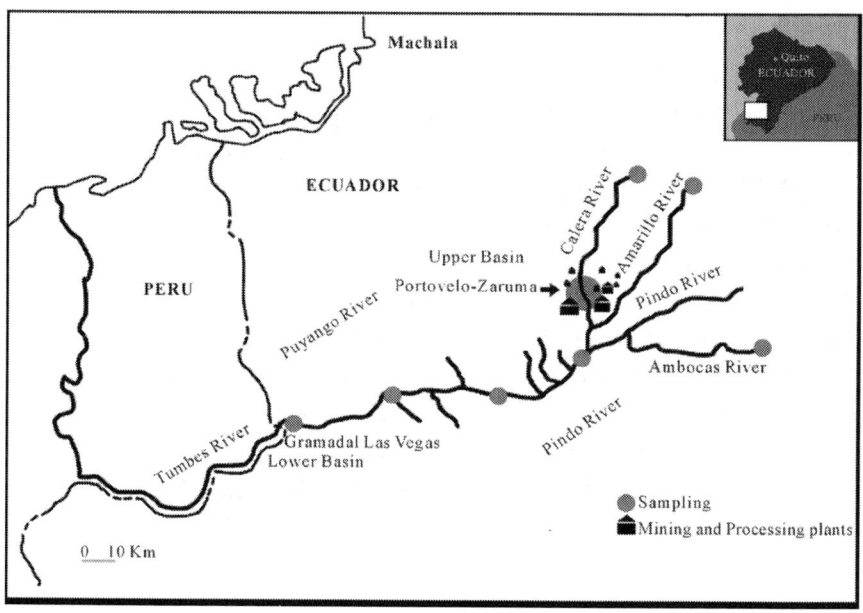

Figure 1: Sampling points of the Puyango River Basin.

Figure 2 shows seasonal variation of pH in both water and sediments. pH values ranging from slightly alkaline to slightly acid that in rainy season is clearly the tendency to neutrality because flux of rivers is increased.

Conductivity has low values at pristine points (areas not affected by the mining pollution), which means a minimum concentration of salts. In contrast, concentration of dissolved salts is notably higher in points with high mining influence, with conductivity that reaches 160 µS/cm. It decreases in most of the tributaries but increases again in the last points (lower Basin), place to which arrive dissolved elements from processing plants area and meddle part of the Basin. There are cations, anions and also heavy metals.

METALLOIDS AND METALS

Manganese

In pristine points (1 to 4), the concentration of manganese in water during rainy season (M_{ed} 1.0 µg/L) is higher than in dry season (M_{ed} 0.6 µg/L). It could be due to effect of dissolution of rocks with natural content of manganese, due to the action of rainfall. In the point 5, a mining area, Mn concentration in winter (63 µg/L) is lower than in summer (159 µg/L). Concentration of manganese are diluted because the increase of precipitation. From 7 to 20 points, in rainy season, increases the concentration of manganese in the main river and tributaries, which may be related to the increased flow that permit mobilization of manganese from the sediments. The most relevant aspect is that in the two stations, the highest levels of manganese found in the processing plant (point 5).

Manganese levels in sediments are higher in dry season, in pristine points, in the area of plants and distal points of the basin. In a similar way for water, manganese, levels reach higher concentrations in the area of the processing plants (1490 µg/g dry season and 809 µg/g rainy season).

The phenomenon is the same in particulate matter per unit of volume; manganese reaches high levels in the mining area (527 g/L in dry season and 197.5 mg/L in rainy season). It is clear that in the last two points, area inhabited by several families, in rainy season there are high levels of Mn (970 µg/L), exceeding even levels of mining area. The volume and force of the flux move the sediments that become part of suspended particulate matter (SPM), material that is abundant in point 20 (408 mg SPM/L, unlike to 140 mg SPM/L in mining area and 3 mg SPM/L as average at pristine points) (Figure 3).

Lead

Lead in water and in pristine points has low concentrations and they are even lower in rainy season (M_{ed} 0.11 µg/L) due to dilution. In contrast, lead concentrations in area of mines are high (30.9 µg/L in dry season and 23.9 µg/L in rainy season). There is a tendency to higher values of lead in lower basin and in the rainy season, probably related to the contribution of lead from sediments and erosion of the Puyango Basin, but these values do not exceed levels of mining area.

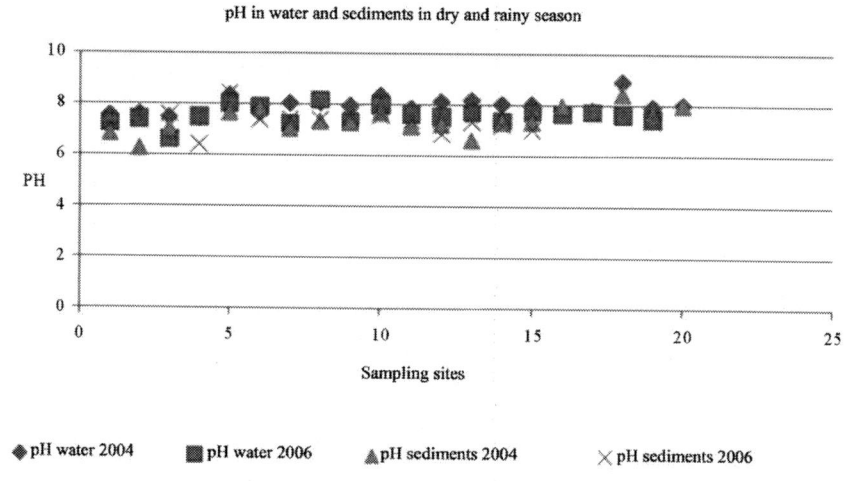

Figure 2: pH in water for seasons, Puyango River Basin.

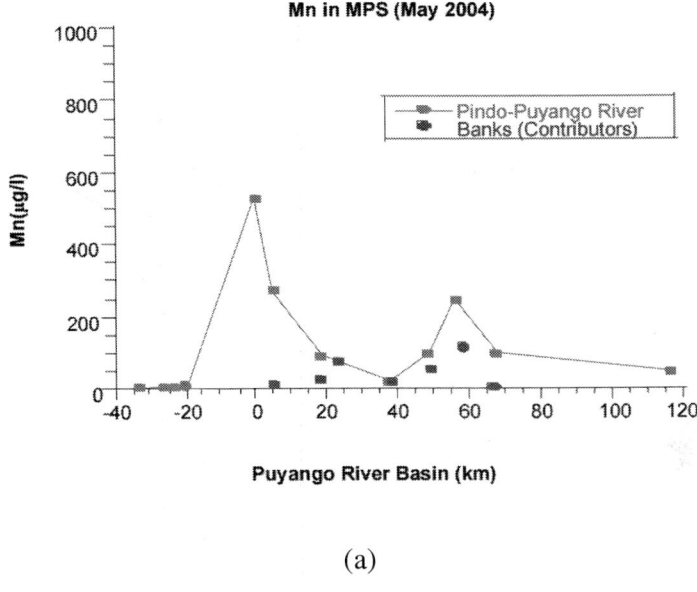

(a)

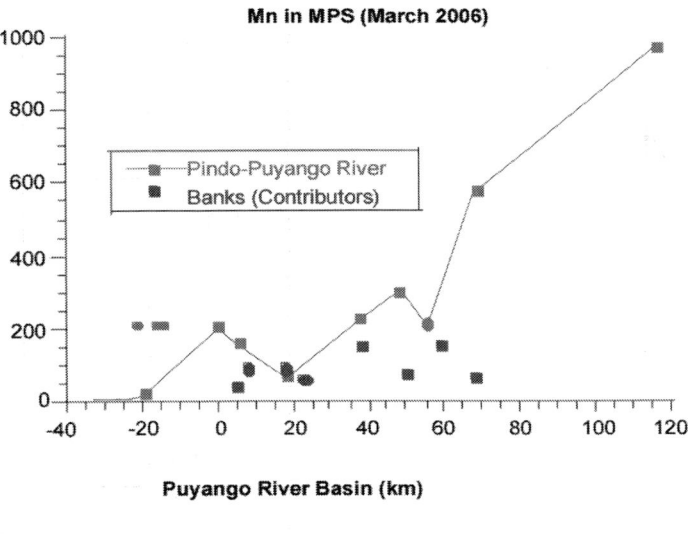

(b)

Figure 3: Manganese in particulate matter per unit of volume, Puyango River Basin.

Values in sediments of pristine points are very low, with slight predominance in dry season (M_{ed} 47.5 µg/g). In contrast, the levels are very high in mining area, in both dry season (440 µg/g) and the rainy season (1090 µg/g), period when the concentrations decreased due to a dilution effect. Lead concentrations in sediments in Lower Basin are high in rainy season (265 µg/g), as well as the values of lead in water, reflecting the dissolution of minerals containing lead in this place.

Lead profile in particulate material is similar to the manganese. Values are very low in pristine points (M_{ed} 0.8 (µg/L in dry season and 0.54 µg/L in rainy season), instead, they increase considerably in mining area, mainly in dry season (648.7 µg/L). In the lower basin, during the rainy season, values increase over the values of the mining area. As we said before, the force of the water influences the presence of particulate matter and metals (Figure 4).

Mercury

Mercury in water and in mining area has high concentrations (7.01 ng/L in rainy season and 2.26 ng/L in dry season), related to Hg in pristine points (M_{ed} 1.46 ng/L in dry season). It is important to remember that gold is amalgamated with mercury in processing plants and waste is eliminated to water bodies. Seasonal variation of mercury is not very significant downstream and the values are lower than mining area.

Mercury in sediments and pristine points has relatively low concentrations (M_{ed} 55 ng/g in dry season and M_{ed} 50 ng/g in rainy season), in contrast, mercury concentrations in mining area are quite high in dry and rainy seasons (820 ng/g and 450 ng/g respectively). Concentrations decrease downstream, however, values are elevated in distal point and in rainy season (220 ng/g). It is the same as other metals (Figure 5).

The mercury profile in the particulate matter is similar to lead and manganese, with the respective elevations in mining area and highest concentrations at the last point in rainy season, generally with low values, below detection limits at various sampling points.

Arsenic

Arsenic concentrations in surface waters in pristine points are similar in both seasons and at very low levels, the highest value is in rainy season (0.63 µg/L). As increases significantly in mining area, more in dry season

(25 µg/l) than in rainy season (5.93 µg/L). Levels at lower basin are similar to mining area, in both dry and rainy seasons (9.0 and 8.41 µg/l respectively), above reference value to drinking water [23] (Figure 6).

As in sediments has high concentrations (161 µg/g) in point 1 (pristine point), source is natural for the presence of arsenopirites in this region, however, points 5, 6, 9 and 12 have the highest concentrations of the basin from mining and mineral dissolution of arsenic. They have high concentrations as 3260 µg/g.

As happen with other chemicals, arsenic concentrations in particulate matter and in mining area are the highest (138 µg/L in dry season). This element is high in rainy season (153 µg/L) because pH is alkaline and this increases the mobility of arsenic from natural sources but also receives arsenic concentrations from mining, deposited in sediments and after mobilized from them.

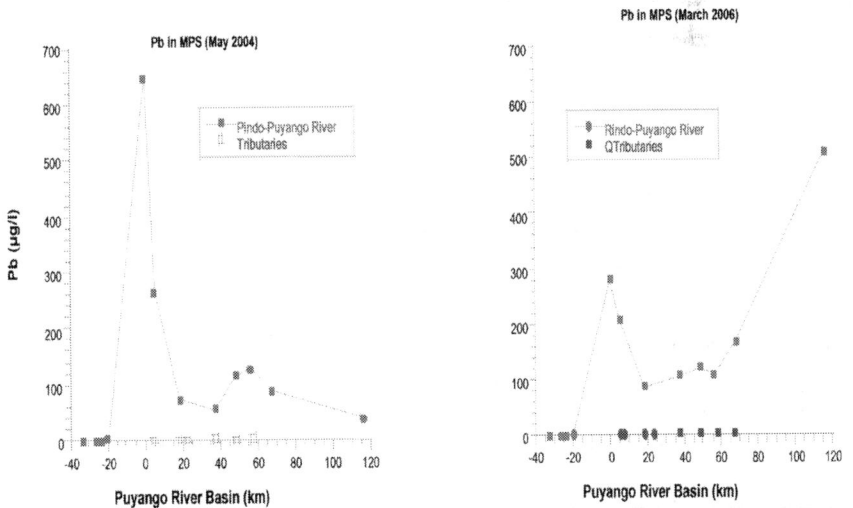

Figure 4: Lead in particulate matter per unit of volume, Puyango River Basin.

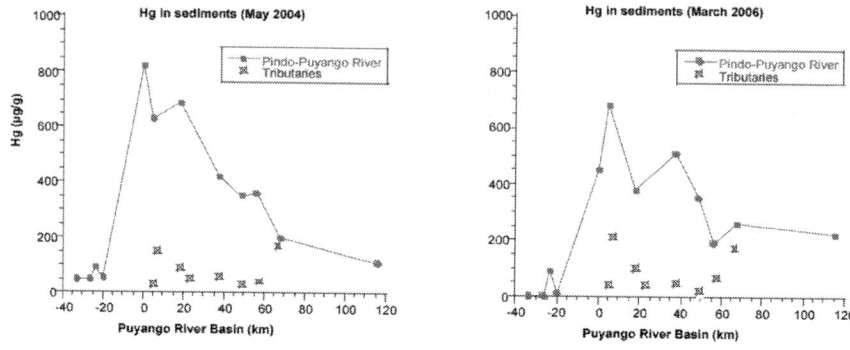

Figure 5: Hg in sediments in the Puyango River Basin (ng/g).

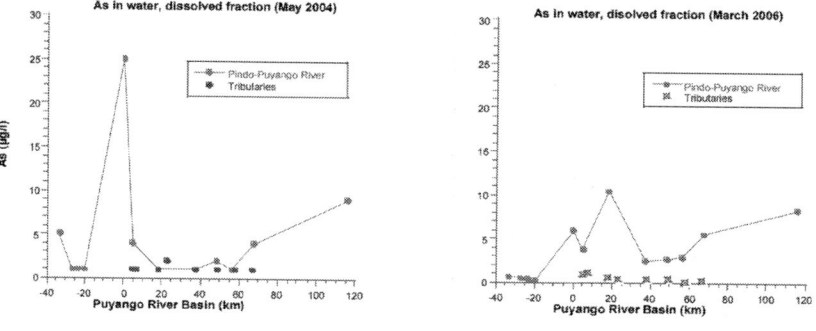

Figure 6: Arsenic in fraction dissolved in the water in Puyango River Basin.

DISCUSSION

Puyango River Basin has a clear difference between pristine points and mining area. The first one presents natural levels o metals; instead, mining area has important anthropogenic pollution. Based on results and with reference to the Ecuadorian legislation contained in the Unified Text of Secondary Legislation of the Ministry of Environment, TULAS [24], Puyango River Basin has high values of lead, manganese and arsenic, which represent a risk to health of the population. This region is usually neutral to slightly alkaline pH with conductivity values showing a low concentration of salts in pristine points and some tributaries. The main problem of this region is an intense mining activity and use of contaminants such as mercury and direct discharge of tailings and effluents from processing plants into the rivers that contribute to high level of these metals. The study shows high values of arsenic, mercury, manganese and lead in points corresponding to mining area. It was found manganese values of 159 µg/L in surface water, in this sector, being 100 µg/L maximum acceptable limit according TULAS. The same happen with lead values, 63 µg/L, being the limit of 50 µg/L. Conditions are better at some points however, most of these chemicals experienced elevation in all substrates at last point corresponding to Gramadal-Las Vegas. Manganese levels reached values of 970 µg/L in rainy season and in this part of the basin, exceeding about ten times the reference values TULAS for drinking water. It is remarkable because in that sector, people consumed water from the river without any treatment, nor filtration. High levels there are with lead (510 µg/L) and arsenic (153 µg/L), being accepted by TULAS limit of 50 µg/L. This contamination has implications for health, for example, in tests on children in this basin, we found that those with higher levels of manganese in their bodies (detected by hair analysis) are children from Gramadal-Las Vegas. In addition, close association was found between levels of manganese and low scores of intellectual abilities of these children (these results are subject of another article). Impact of manganese on human health, more than 115 miles from source of contamination is an example of how mining impacts far away from it, according to the habits, origin and use of resources that population consumes. Population up stream knows that river water is not good to drink. They have other alternatives.

It is also significant the variation in the concentrations of toxic elements between the rainy and dry season.

A particular phenomenon happens with mercury concentrations in water that do not exceed 7 ng/L in mining area and 4.25 ng/L in lower basin. In contrast, there is a large difference in the levels of mercury in sediments comparing values of pristine points and that of mining area. Concentrations are not more than 55 ng/g, instead, in mining area reach values of 820 ng/g and in the lower basin reaches 220 ng/g. These findings may suggest that mercury transforms into organic mercury (MeHg) and pass the food chain up to the human body, however, the study found almost absence of methylmercury in the main axis of Puyango River, phenomenon attributed to low bacterial activity in its ability to methylate mercury, which is associated clearly with the concentrations of cyanide, pollution from mining, details of this interesting phenomenon are found in an article [8].

"Mercury, manganese and lead, at toxic levels, inhibit the functioning of the nervous system. Neurobehavioural tests are used to assess motor, sensory, and cognitive functions. Poor performance on these tests has been linked to metal toxicity. We get some impacts of heavy metals in other component of our study. In 229 adult men working in, or living near, the mining areas, elevated blood mercury and lead seemed to be associated with poor performance on neurobehavioural tests. Increased concentrations of manganese in the hair of girls (2.9 - 7.4 mg/g) were associated with decreased scores on the cognitive Raven test ($p = 0.009$) and the digits test ($p = 0.03$). In children, increased concentrations of mercury in hair (0.1 - 4.3 mg/g) was associated with decreased performance on Santa Ana dexterity ($p = 0.005$), digits ($p = 0.01$), and finger tapping ($p = 0.04$) tests.

The values of mercury in the Puyango River Basin shown in some places higher values than studies in an urbanized Mediterranean area, which main source of contamination is the industrie, with concentrations of Hg and MeHg in sediments between 0.04 and 0.64 µg/g [25].

Comparing the values of mercury with other regions such as China, which also has mining regions that extract gold using mercury, river water showed concentrations of this element of 1.0 ng/l (Puyango basin also shows higher values surface water) and sediments from 100 to 300 ng/kg [26].

Some studies in the Amazon region showed that a major source of mercury pollution in the atmosphere is burning of the gold-mercury amalgam, which releases into the atmosphere from 30 to 170 tons of mercury per years. Atmosphere emissions account for 45% and 87% of the total mercury emitted by the amalgamation and burning of gold.

Other studies in the Brazilian Amazon show absence a clear profile of mercury emissions mercury from gold mining areas, whereas high concentrations were found in natural soil. Erosion and deforestation are the main source of mercury deposited in aquatic systems [27-29]. In contrast, this study shows mining impact throughout the entire basin and throughout the year.

A study in northeastern Argentina, site that has a history of having high concentrations of arsenic, determined that all sampling values exceeded the maximum allowed under the Law of Argentina (0.05 mg/L) [30]. This study showed that pollution comes from natural sources such as in pristine points the Puyango River Basin.

Lead was found at elevated levels in sites with direct influence of mining activities, accumulated in sediments and decreasing in lower basin. Values are not significantly different in the dry season and the rainy season, which indicates no significant dissolution of this element in the rainy season because pH does not allow the dissolution of lead salts. There is no sudden change of pH in the entire basin.

The results have helped to take some protective measures in lower river basin of Puyango, such as filtration systems and use of local sources to obtain dinking water. It was found high values of mercury (0.1 to 0.5 mg/L) in the Lake Vaner of Sweden, one of the reservoirs largest freshwater in this country and significantly affected by mercury pollution for more than fifty years. Mercury was removed almost entirely after applying remediation processes [31]. By contrast, in rivers like Puyango, with intense mining activity, reducing the metal contamination depends of profound changes in mining practice. In the upper basin, we can get some institutional policies to control the pollution from mining [32] (creation of operating units of study and control, municipal bylaws to prevent anarchy and polluting work of processing plants and greater control with the Ministry of Environment and not Renewable Natural Resources).

ACKNOWLEDGEMENTS

This study was financially supported by the International Development Research Center (IDRC-CRDI), Ottawa, Canada, www.idrc.ca.

REFERENCES

1.   M. Cortazar, "El Oro de Portovelo," MC Editors, Soboc Grafic, Quito, 2005, p. 152.

2.   T. Bustamante and R. Lara, "El Dorado o la Caja de Pandora, Matices Para Pensar la Minería en Ecuador," Flacso Sede Ecuador, Quito, 2010, p. 145.

3.   M. E. Garcia, et al., "History of Mining around the Poopo Lake and Environmental Consequences," Vatten, Vol. 61, 2005, pp. 243-248.

4.   R. Paredes, "Oro y Sangre en Portovelo, el Imperialismo en el Ecuador," Artes Gráficas, Quito, 1938, p. 228.

5.   A. Acosta, "La Maldición de la Abundancia," CEP-Abya Yala, 1ra Edición, Quito, 2009, p. 239.

6.   O. Betancourt, "Para la Enseñanza e Investigación de la Salud y Seguridad en el Trabajo," FUNSAD-OPS/OMS, Primera Edición, 1999, pp. 171-232.

7.   M. Priester and T. Hentschel, "Small-Scalle Gold-Mining Processing Techniques in Developing Countries," GATEGTZ, Vieweg, 1992, pp. 15-81.

8.   J. R. D. Guimaraes, O. Betancourt, R. Barriga, E. Cueva and S. Betancourt, "Long-Range Effect of Cyanide on Mercury Methylation in a Gold Mining Area in Southern Ecuador," Science of the Total Environment, Vol. 409, No. 23, 2011, pp. 5026-5033. doi:10.1016/j.scitotenv.2011.08.021

9.   F. Hruschka and C. Salinas, "Estudio Colectivo de Impacto Ambiental y Plan de Manejo Ambiental para las Plantas de Beneficio Mineral Aurífero ubicadas en la Vega del Río Calera/Salado," CENDA-COSUDE, Projekt Consult, Mimeo, Zaruma, 1996.

10.   Prodeminca, "Evaluation of Impacts in the Mining District of Zaruma-Portovelo and the Puyango River Basin. Proyecto Desarrollo Minero y Control Ambiental, Prodeminca, Swedish Environmental Systems," Ministerio de Energia y Minas del Ecuador, Quito, 1998, p. 212.

11.   N. H. Tarras-Wahlberg, A. Flachier, S. N. Lane and O. Sangfors, "Environmental Impacts and Metal Exposure of Aquatic Ecosystems in Rivers Contaminated by Small Scale Gold Mining: The Puyango River Basin, Southern Ecuador," Sciences of the Total Environment, Vol. 278, No. 1-3, 2001, pp. 239-261. doi:10.1016/S0048-9697(01)00655-6

12.   P. C. Velasquez-Lopez, M. M. Veiga and K. Hall, "Mercury Balance in Amalgamation in Artisanal and SmallScale Gold Mining: Identifying Strategies for Reducing Environmental

Pollution in Portovelo-Zaruma, Ecuador," Journal of Clean Production, Vol. 18, No. 3, 2010, pp. 226-232. doi:10.1016/j.jclepro.2009.10.010

13.   J. Marrugo, L. Benitez and J. Olivero, "Distribution of Mercury in Several Environmental Compartments in an Aquatic Ecosystem Impacted by Gold Mining in Northern Colombia," Arch Environ Contam Toxicol, Vol. 55, 2008, pp. 305-316. doi:10.1007/s00244-007-9129-7

14.   L. M. Bourgoin, et al., "Mercury Pollution Due to Gold Mining in the Bolivian Amazonian Basin. Mercury as a Global Pollutant," 5th International Conference, Río de Janeiro, 1999, p .152.

15.   Bravo, et al., "Assessment of Mercury Levels in Soils, Waters, Bottom Sediments and Fishes of Acre State in Brazilian Amazon," Water, Air, and Soil Pollution, Vol. 147, No. 1-4, 2003, pp. 61-77.

16   R. Cesar, et al., "Mercury, Copper and Zinc Contamination in Soils and Fluvial Sediments from an Abandoned Gold Mining Area in Southern Minas Gerais State, Brazil," Environmental Earth Science, Vol. 64, No. 1, 2011, pp. 211-222.

17.   J. R. D. Guimaraes, O. Malm and M. Meili, "Mercury in Soils, Sediments and Fish around the Poconé Gold Mining Area, Pantanal, Brazil: Some Movilisation but No Health Risks. Mercury as a Global Pollulant," 5th International Conference, Río de Janeiro, 1999, p. 154.

18.   J. Howard, et al., "Total Mercury Loadings in Sediment from Gold Mining and Conservation Areas in Guyana," Environ Monit Assess, Vol. 179, No. 1-4, 2011, pp. 555- 573.doi:10.1007/s10661-010-1762-3

19.   S. Montgomery, et al., "Total Dissolved Mercury in the Water Column of Several Natural and Artificial Systems of Northern Quebec Canada," Canadian Journal of Fisheries and Aquatic Sciences, Vol. 52, No. 11, 1995, pp. 2483-2492.

20.   M. Roulet, et al., "Mercury and Other Trace Metals Dispersion from Gold Mines in the Puyango River, Ecuadorian Andes," 6th International Conference on Mercury as a Global Pollulant, Japan, October 2001.

21.   O. Betancourt, N. Alberto and R. Marc, "Small-Scale Gold Mining in the Puyango River Basin, Southern Ecuador: A Study of

Environmental Impacts and Human Exposures," EcoHealth, Vol. 2, No. 4, 2005, pp. 323-332. doi:10.1007/s10393-005-8462-4

22. O. Betancourt, et al., "Environmental and Health Impacts of Small Scale Gold Mining in Ecuador (Phase 2)," Informe Técnico, FUNSAD, 2007.

23. ATSDR, "Tox Guide for Arsenic As," CAS 7440-38-2, 2005. www. atsdr.cdc.gov.toxiguides/toxguide-2.pdf

24. Gobierno del Ecuador, "Texto Unificado de Legislación Secundaria del Ministerio del Ambiente, Libro VI, Anexo 1, Norma de Calidad Ambiental y de Descarga de Efluentes: Recurso Agua," Registr Oficial, Edición Especial 2, 2003.

25. C. Abi-Ghanem, "Mercury Distribution and Methylmercury Mobility in the Sediments of Three Sites on the Lebanese Coast, Eastern Mediterranean," Achieves of Environmental Contamination Toxicology, Vol. 60, 2010, pp. 394-405.

26. Y. H. Lin, M. X. Guo and W. M. Gan, "Mercury Pollution from Small Gold Mines in China," Water, Air, and Soil Pollution, Vol. 97, No. 3-4, 1995, pp. 233-239.

27. M. Roulet, M. Lucotte, R. Canuel, N. Farella, M. Courcelles, J. R. D. Guimarães, D. Mergler and M. Amorim, "Increase in Mercury Contamination Recorded in Lacustrine Sediments Following Deforestation in Central Amazonia," Chemical Geology, Vol. 165, No. 3-4, 2000, pp. 243-266. doi:10.1016/S0009-2541(99)00172-2

28. M. Roulet, M. Lucotte, N. Farella, G. Serique, H. Coelho, C. J. Passos, E. D. De Jesus Da Silva, P. S. de Andrade, D. Mergler, J. R. D. Guimaraes and M. Amorim, "Effects of Recent Human Colonization on the Presence of Mercury in Amazonian Ecosystems," Water, Air and Soil Pollution, Vol. 112, No. 3-4, 1999, pp. 297-313. doi:10.1023/A:1005073432015

29. M. Roulet, M. Lucotte, A. Saint-Aubin, S. Tran, I. Rhéault, N. Farella, E. D. De Jesus Da Silva, J. Dezencourt, C. J. Sousa Passos, G. Santos Soares, J. R. D. Guimarães, D. Mergler and M. Amorim, "The Geochemistry of Hg in Central Amazonian Soils Developed on the Alter-DoChão Formation of the Lower Tapajós River Valley, Pará State, Brazil," The Science of the Total Environment, Vol. 223, No. 1, 1998, pp. 1-24. doi:10.1016/S0048-9697(98)00265-4

30. P. Motta, "Hierro y Manganeso en Aguas Superficiales y Subterráneas," Argentina, 2005.

31. A. Danielsson, "A Large Scale Mercury Variation in Lake Vaner, Sweden, 2001.

32. O. Betancourt, et al., "Impacts on Environmental Health of Small-Scale Gold Mining in Ecuador," In: D. F. Charron, Ed., Ecohealth Research in Practice: Innovative Applications of an Ecosystem Approach to Health, Springer, New York, 2012.

Chapter 4

Geomatics for Rehabilitation of Mining Area in Mahis, Jordan

Rami Al-Ruzouq[1] and Samih Al Rawashdeh[2]

[1]Department of Civil and Environmental Engineering, University of Sharjah, Sharjah, UAE

[2]Department of Surveying and Geomatics Engineering, Al-Balqa Applied University, Al-Salt, Jordan

ABSTRACT

Mining activities often cause dramatic changes in landscapes, particularly in the dump sites and its surrounding environment. Land rehabilitation is the process of renovating damaged land to some extent of its original shape and aims to minimize and mitigate the environmental effects to allow new land uses. The success of different rehabilitation strategy and newly suggested urban and architecture modeling depends on the landscape characterization (topography of the study area and its derivatives such as slope and aspects, geological and geomorphologic nature of the study area). The aim of this study is to demonstrate the utility of different methodologies based on geomatics techniques (Photogrammetry, Remote Sensing, Global Positioning System (GPS) and three dimensional Geographic Information System (GIS)) for highlighting landscape characterization which is needed for rehabilitation of Mahis area. Photogrammetric adjustment procedures were used to create digital elevation model and Orth-Photo model for the study area using aerial images. Remote sensing data were used for land classification to provide vital information for rehabilitation planning. GPS field observations were used to build spatial network for the study area based on ground control point collections. Finally, realistic representation of the study area with three dimensional GIS was prepared for the study area considering ease and flexible updating of the geo-spatial database.

INTRODUCTION

Rehabilitation is the process of returning the land in a given area to some degree of its former self, after some process (business, industry, natural disaster etc.) has damaged it [1] . Many projects and developments will result in the land becoming degraded, for example mining, farming and forestry. Different Geomatics techniques were used to prepare the comprehensive knowledge base needed for the rehabilitation process; first we started with collecting the Ground Control Points by using GPS, second, remote sensing techniques were applied such as image classification and Band Fusion. Photogrammetric techniques were used to produce DTM and Ortho-photo for the study area. Moreover, 3D modeling using AutoCAD was used to add different modules to the Orthophoto. Finally by using GIS, different layers like (contour lines and Faults) were added.

Mahis mining study area is about 250 m² and located at Latitude: 31°58›60N, Longitude: 35°46›0E and 810 meters above mean sea level, Figure 1. It is between Mahis and Al-Fuhays towns, which are connected with Al-Salt, and the capital Amman about 30 km away. It is also about 35 km from the Dead Sea which is lowest spot located on the earth. For the past 30 years, this mine has been producing some of the highest grade Kaolins in the area where the output capacity reaches more than 70,000 tons annually. Figure 2 shows actual situation of the Mahis quarry.

Literature shows that many projects that dealing with rehabilitation of opencast mining area using Geomatics. Most of these projects deal with Photogrammetry Global positioning system (GPS), remote sensing, geophysics and GIS to prepare and organize the geo-base knowledge of rehabilitation process. The legacy of an estimated twenty-seven thousand abandoned mines pose environmental, health, safety and economic problems to communities, the mining industry and governments across Canada [2] . Information gathering for these sites is necessary to enable sound decision-making, cost-efficient planning and sustainable rehabilitation. The sustainable management and rehabilitation of mine sites for decision support project is working collaboratively with federal departments, provincial governments and industry to develop new techniques for information collection and integration to support mine reclamation and policy decisions surrounding mine rehabilitation.

In support of the Ontario Ministry of Mines and Northern Development (OMND)'s Abandoned Mine Rehabilitation Program, scientists are providing a case study using new techniques at the KamKotia mine in northern Ontario. Copper and zinc were mined at KamKotia from 1942-1972, before it was abandoned and left under government stewardship. During this period, three million tons of tailings were discharged in the area around the mine, before an impoundment was built in 1967. This produced a "kill zone" of approximately 170 hectares [3].

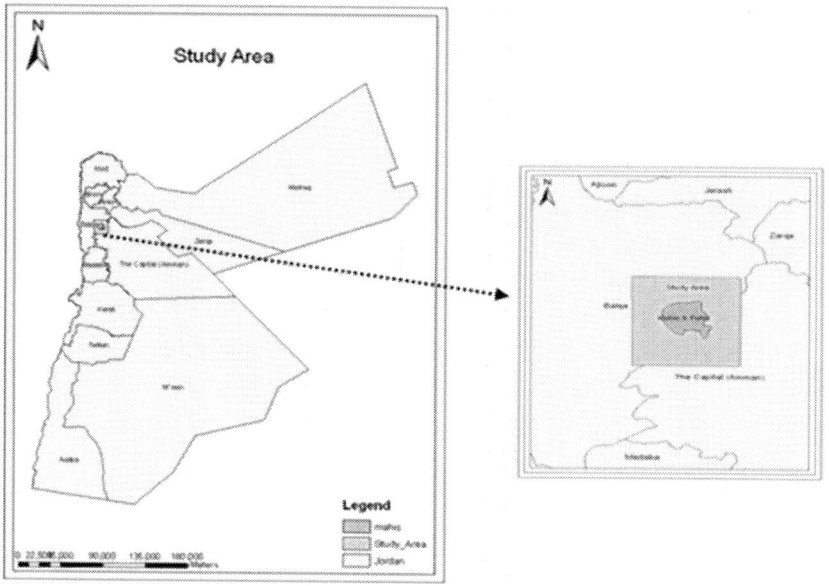

Figure 1: Location of the study area.

Figure 2: Actual situation of the mahis quarry.

In 2001, the OMND began a multi-year project to rehabilitate the site. To date, a water-treatment plant and a new tailings impoundment have been built, and the exposed tailings areas are being dredged and deposited in the new impoundment. Baseline map of the site was developed using information extraction techniques for remote sensing data, developed in collaboration with the Geo-research Centre (GFZ) in Germany. The map was derived from airborne "hyper-spectral" remote sensing data collected in 2001, with a five-meter spatial resolution. Hyper-spectral sensors collect reflected radiation from the earth's surface in a large number of narrow spectral bands. In 1991 and 1998, the International Aluminum Institute (IAI) commissioned surveys regarding bauxite mine rehabilitation programs that had been undertaken by operations around the world. The aim in both cases was to provide data on the environmental impacts of bauxite mines and their rehabilitation programs. In 2003, a third survey was carried out to follow up and extend the first two. The survey shows that bauxite miners are making substantial efforts towards the sustainable development of the industry. Moreover, this project includes items that are not founded in the previous rehabilitation projects such as: different layers cover the area (contour lines, faults, hospitals…), bands combinations, DTM and Orthophoto generation, and 3D modeling [4] [5] .

The objectives of this project can be summarized by two main points: first, create geo-database of the study area (Mahis, Jordan). To evaluate the rehabilitation process for the opencast mining area. The Geodatabase include: Digital Terrain Model DTM from different sources such as SPOT, Contour lines and aerial photos, Ortho-photo, Land use map, Ground Control Points and Structural map and finally Produce a three-dimensional Modeling GIS for the study area; second, improvement of the ecological, economic, and social situation of the region, to satisfy the needs of today's generation without threatening the quality of life of the coming generations.

In this paper, we will discuss the various steps for spatial data collection and processing in the Mahis area to establish the geographic information system. This includes, photogrammetry processing based on aerial and satellite discussed in Section 2. Section 3 illustrates the output of remote sensing based on Landsat satellite image. Section 4 describes various elements of Geographic Information System (GIS). Three dimensional modelling of produced geospatial element were depicted in Section 5 for better visualization and interpretation capabilities. Rehabilitation process and requirements needed for the suggested modeling were discussed in Section 6. Finally, conclusion and future work are discussed in Section 7.

PHOTOGRAMMETRIC PROCESSING

Photogrammetry was the main source of spatial data in our project, two main photogrammetric products generated; ortho-photo, and digital elevation model (DEM). As shown in Figure 3, stereo aerial photos were used in this project and have the following properties:

- Aerial photos scanned from hard-copy images and one strip directed (North-South).
- 60% overlap between two adjacent photos.
- 1:25,000 photos scale and 23 cm^2 film format.
- Spatial resolution is about 50 cm^2 with 5.7 km^2 coverage per photo.

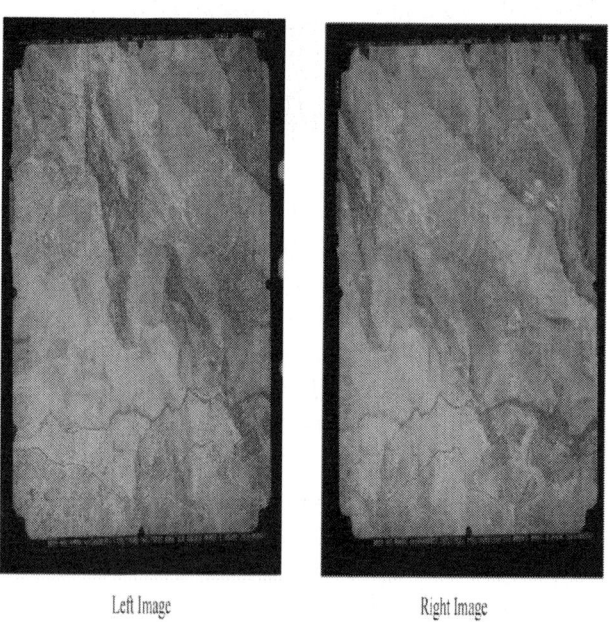

Left Image   Right Image

Figure 3: Aerial photos for Mahis area (Jordan).

Digital Terrain Model (DTM) Extraction

The concept of creating digital models of the terrain is relatively recent development. DTM is simply a statistical representation of the continuous surface of the ground by a large number of selected points with known X, Y, Z coordinates in an arbitrary coordinate field. The choice of data sources and terrain data sampling techniques is critical for the quality of the resulting DTM. At present, most DTM data are derived from three alternatives source: Ground surveys, Photogrammetric data capture (which is the data source of the DTM in this project), and Digitized cartographic data sources. Figure 4 shows the resultant DTM after aerial triangulation process for the aerial photos using ground control points described in the next Section. Figure 4 shows the variation in elevation from 396 - 736 m above mean sea level where white color shows highest elevation [5] [6] .

Ortho-Photo Generation

An ortho-photo is photo that has the same characteristics of a map. Thus, ortho-photo can be used as maps to make measurements and establish accurate geographic location of features. Ortho-photos are generated from aerial photographs and satellite images through a process known as ortho-rectification. A normal (uncertified) aerial photograph and satellite images does not show features in their correct locations due to displacements caused by the tilt of the sensor and terrain. Ortho-rectification transforms the central projection of the photograph into an orthogonal view of the ground. Therefore it is removing the distorting effects of tilt and terrain [7].

Generation of an ortho-photo map from aerial photograph requires information on the location of the camera and its orientation in space as well as a model of the terrain elevation. In this project, ortho-photo was generated for the study area using information obtained from the photogrammetric processing and the DTM. Figure 5 shows the three dimensional result of the Ortho-photo and DTM.

Ground Control Points (GCPs)

Photogrammetric control consists of any points whose positions are known in an object-space reference coordinate system and whose images can be positively identified in the photographs. In aerial photogrammetry, the object space is the ground surface, and various reference ground coordinate systems are used to describe control point positions. Photogrammetric control, or ground control as it is commonly called in aerial photogrammetry, provides the means for orienting or relating aerial photographs to the ground [8] .

After collecting the project aerial-photos, GCPs is needed. Thus, planning and distribution of the GCPs on the photos (Planning Stage) is necessary in order to make the suitable selection of GCP which covered study area, as in the following Figure 6.

Difficulties during GCP collection due to the steep and high mountains topography of the study area is where only very few interest and sharp features appearing in the aerial and it was not easy to be distinguished.

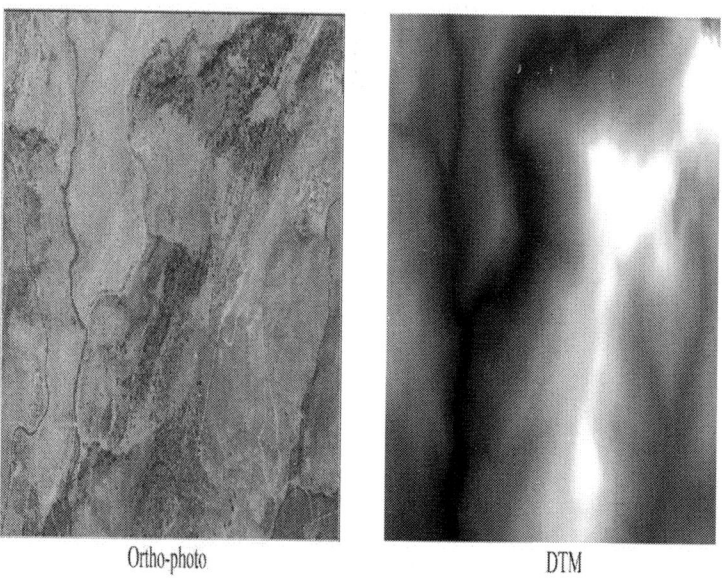

Ortho-photo   DTM

Figure 4: Ortho-Photo and DTM generated after aerial triangulation process.

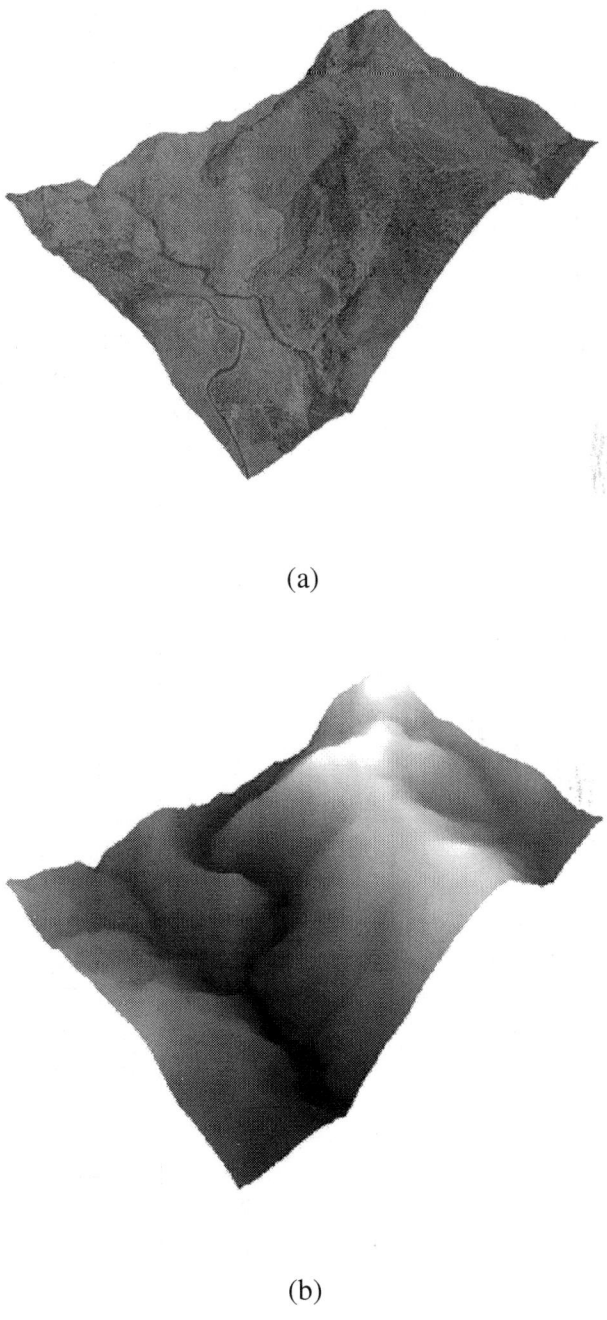

(a)

(b)

Figure 5: The Ortho-photo and the DTM in 3D view.

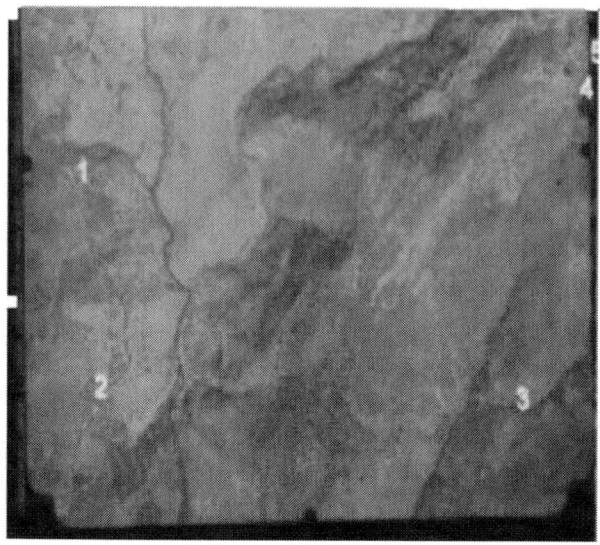

Figure 6: GCPs distribution in the overlap region.

REMOTE SENSING

Remote sensing is concerned with acquiring spatial information from a range of sensors, including satellite imagery, airborne scanners and radar satellites. Remote sensing provides important coverage, mapping and classification of land cover features, such as vegetation, soil, water and forests. In this project, remote sensing focuses on Band-Combination and classification. The data used for this purpose was Landsat Scenes covering Jordan and some of the nearby countries. The Landsat scenes were in the 6 TM bands (bands 1, 2, 3, 4, 5, and 7), Figure 7. The scenes are geo-referenced images in the UTM36N WGS84 coordinate system. The spatial resolution of these scenes is 30 m and captured in year 2000.

Supervised Classifications

Supervised classification is more closely controlled by you than unsupervised classification it requires more input and experience by the analyst but it can produce more accurate and useful results than unsupervised classification. In this process, you select recognizable regions within an image, with help from other sources, to create sample areas called training sites. Your training sites are then used to train the computer system to identify pixels with similar characteristics. Knowledge of the data, the classes desired, and the algorithm to be used, is required before you begin selecting your training sites. By setting priorities to your classes, you supervise the classification of pixels as they are assigned to a class value [9] [10] .

Maximum likelihood Classification is a statistical decision criterion to assist in the classification of overlapping signatures; pixels are assigned to the class of highest probability. The maximum likelihood classifier is considered to give more accurate results than parallelepiped classification however it is much slower due to extra computations, Figure 8.

Unsupervised Classification

An unsupervised classification organizes image information into discrete classes of spectrally similar pixel values [9]. This is a highly computer-automated procedure. In an unsupervised classification the software automatically divides the range of spectral values, contained in an image file, into classes. A classification report can indicate the presence of a specific ground cover because a proportion of the classified pixels fall within its known spectral signature.

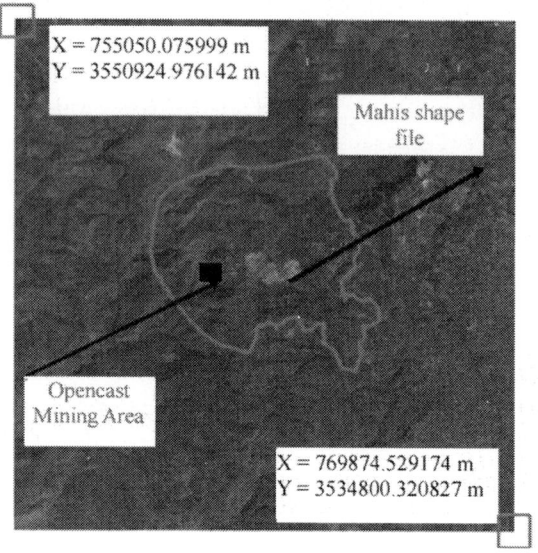

Figure 7: Land-sat subset image for the study area.

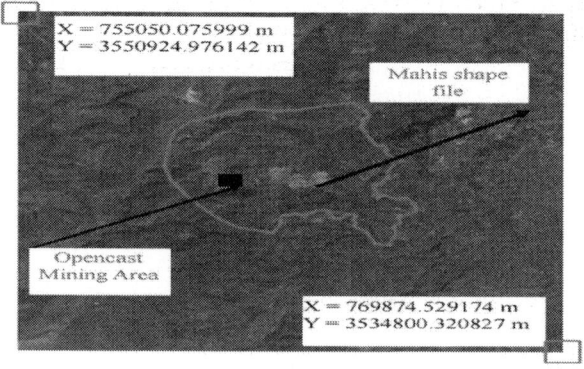

Figure 8: Maximum likelihood classification.

In such a case, you need to know what the spectral signature of the target ground cover is in order to identify its presence. Unsupervised learning or clustering is a way to form "natural groupings" or clusters of patterns. K-Means method works by choosing random seeds, which can be thought of as points with random DN values. After the seeds have been chosen lines are formed to separate the classes. Next, the points lying within the delineated areas are analyzed, and their means are noted. The means then form the new seeds, and a new series of lines are formed to separate the classes. This process is then repeated several times, Figure 9. Assuming that the number of clusters is known and well defined, k-means algorithm has been adopted rather than the ISODATA algorithm.

Bands Combination

Satellites acquire images in black and white, and it is possible to create the beautiful color images. Images created using different bands (or wavelengths) have different contrast (light and dark areas) [9] [11]. Computers make it possible to assign "false color" to these black and white images. The three primary colors of light are red, green, and blue; computer screens can display an image in three different bands at a time, by using a different primary color for each band. Different combination for these three images we get a "false color image".

Combination 3, 2, 1: the "natural color" band combination, Figure 10. The visible bands are used in this combination, ground features appear in colors similar to their appearance to the human visual system, healthy vegetation is green, recently cleared fields are very light, unhealthy vegetation is brown and yellow, roads are gray, and shorelines are white. This band combination provides the most water penetration and superior sediment and bathymetric information. It is also used for urban studies.

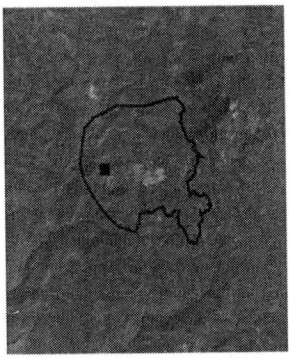 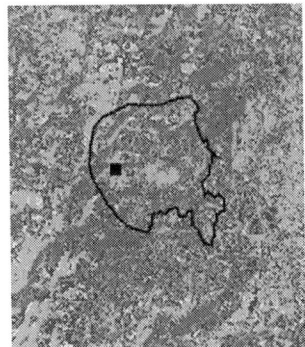

Figure 9: K-means classification.

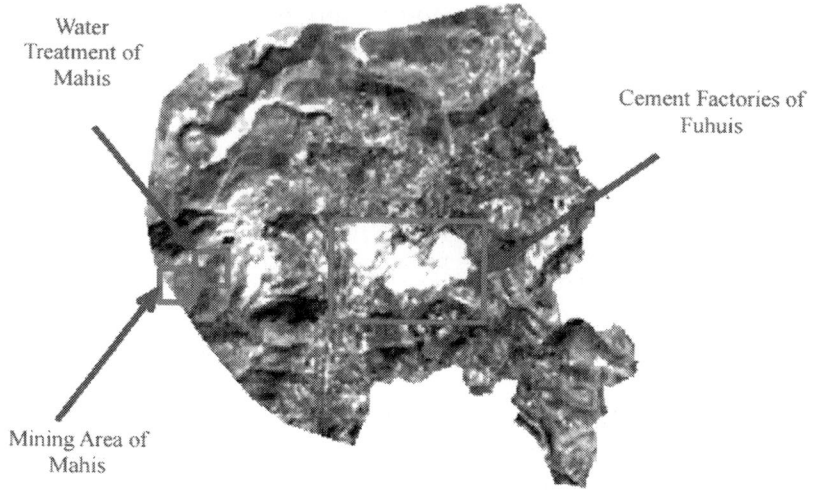

Water
Treatment of
Mahis

Cement Factories of
Fuhuis

Mining Area of
Mahis

Figure 10: Mahis and fuhuis from space (321 ETM+).

Combination 4, 3, 2: the standard "false color" composite. Vegetation appears in shades of red, urban areas are cyan blue, and soils vary from dark to light browns. Ice, snow and clouds are white or light cyan. Coniferous trees will appear darker red than hardwoods. This is a very popular band combination and is useful for vegetation studies, monitoring drainage and soil patterns and various stages of crop growth. Generally, deep red hues indicate broad leaf and/or healthier vegetation while lighter reds signify grasslands or sparsely vegetated areas. Densely populated urban areas are shown in light blue. This TM band combination gives results similar to traditional color infrared aerial photography, Figure 11.

GEOGRAPHIC INFORMATION SYSTEM (GIS)

In general, data entry can be very time consuming, but it is the most important task of the GIS process. This Section discusses the basic organization of entering data, scanning, layers designing, digitizing, geo-referencing and projection, (3D) applications in GIS, creating a layout for study area [12]. In this paper, "ArcGIS" program was used in the GIS processing that include scanning, geo-referencing and digitizing,

Maps Scanning and Geo-Referencing

A Map can be defined as extremely accurate sketch which simulate the reality with two fundamentals special effects. The first is the map scale which is the ratio between distances on the map to the same distance on ground. The second is the map projection and the coordinate system which is defined as systematic transformation of the spheroidal shape of the earth so that the curved, three-dimensional shape of a geographic area on the earth can be represented in two dimensions, as x, y coordinates. Projection formulas are mathematical expressions that convert data from a geographical location on a sphere or spheroid to a representative location on a flat surface. See Table 1 and Figure 12 which shows a sample of the processed topographic and geological map.

Figure 11: 4, 3, 2 band combinations.

Table 1: List of the scanned topography and geology maps.

Name	year	Type	Scale
Mahis	1992	Topography	1:10,000
Wadi as ser	1992	Topography	1:10,000
Sweileh	1991	Topography	1:50,000
Dead sea	1997	Topography	1:50,000
Amman	2001	Geology	1:50,000
Salt	1993	Geology	1:50,000
Sweileh	1993	Geology	1:50,000
Dead sea	2001	Geology	1:50,000

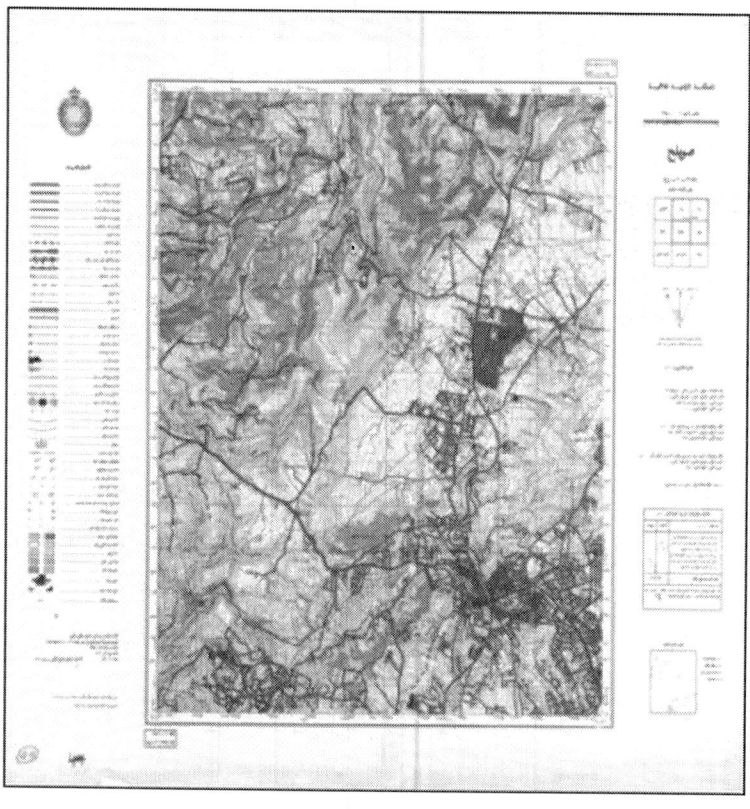

(a)

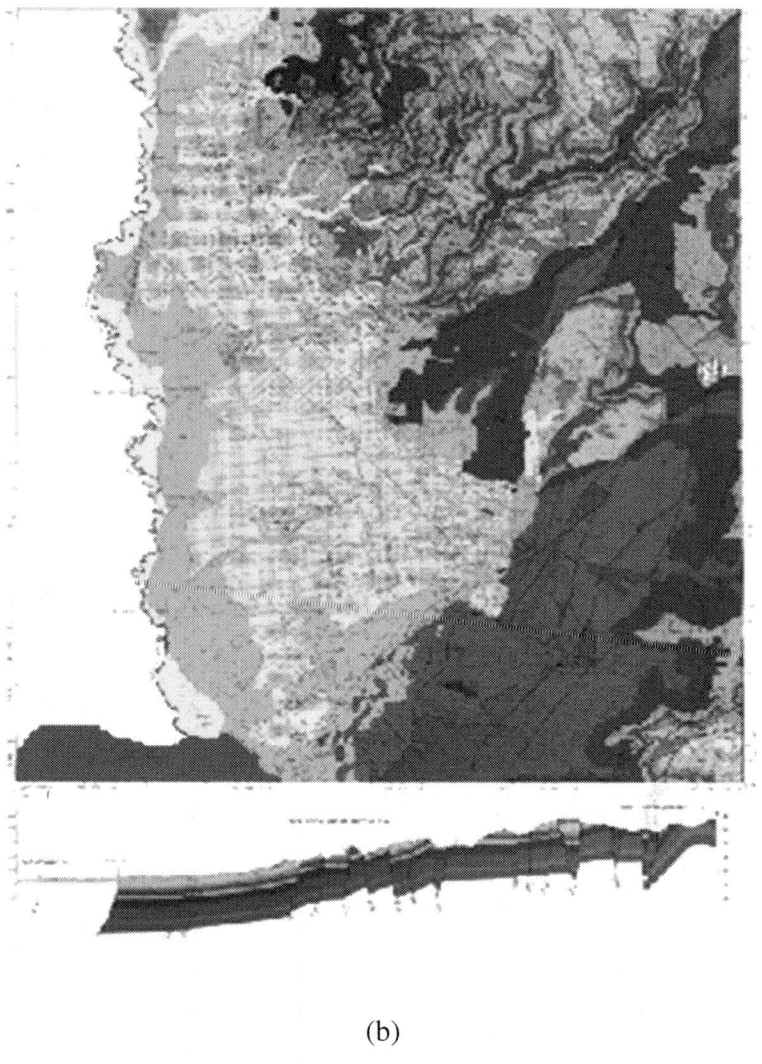

(b)

Figure 12: Sample of scanned maps. (a) Topography; (b) Geology.

Structural Map

Structural Map contains many structural elements such as faults, rocks composite, wells to help us lately to select the proper sites for buildings, farming, etc. [13] . The study include four geological maps [Amman,

AS-Salt, Suwaylih, Al-Karama]. Digitized Faults from Geology Map with scale 1:50,000 shown in Figure 13.

THREE DIMENSIONAL MODELLING

A triangulated irregular network (TIN) is a digital data structure used in a geographic information system (GIS) for the representation of a surface. A TIN is a vector-based representation of the physical land surface or sea bottom, made up of irregularly distributed nodes and lines with three-dimensional coordinates (x, y, and z) that are arranged in a network of no overlapping triangles. TINs are often derived from the elevation data of a rasterized digital elevation model (DEM) [14].

Triangles in a TIN can are oriented in different directions. The orientation of triangles is referred to as triangulation. Figure 14 shows some triangles that are oriented vertically while others are oriented horizontally. Thereafter, the TIN can is converted to raster with 10m resolution. The slope and aspect (direction of slope) has been also derived from TIN. Creating the Slope and Aspect are very important in our rehabilitation process, to know exactly where to build the suggested modules, Figure 14.

CAD Modeling

CAD (Computer Aided Design) files, can be displayed in GIS. Many CAD files are created in a local coordinate system, and must be transformed or moved, to align with GIS data in a real-world coordinate system [15]. Figure 15(a) explores the Geo-referencing 3D-Models above ortho-photo while Figure 15(b) shows the rendered 3D-Model after Transformation.

REHABILITATION

A Rehabilitation Plan for abandoned mine reclamation should contain certain key elements [16]. These elements are important whether the concern is mine fire control, mine subsidence prevention, mine hazard removal or mine drainage abatement. The elements are inter-related with information from one element feeding the others. The development of rehabilitation plans is an evolutionary process. Plans begin with a vision and move forward through the initiative, commitment and perseverance of the involved partners. It is not something that can be put together in a week or month. In the beginning, the content of the plan and each element in the plan may be more conceptual than real. As work is completed, the focus will become clearer and the plan will take on some substance. A Rehabilitation Plan should include the following elements: Goals must be reasonable and achievable. There should be a deadline for achieving the established goals. The time schedule will be used to develop the financing plan. Goals can be short-term and/or long-term. The benefits to be gained by achieving the goals should be thoroughly discussed. Technical alternatives for addressing the problems, including the costs, must be considered. The alternatives should identify both conventional technologies and innovative technologies that reduce the cost of reclamation. The pros and cons of the alternatives should be discussed. The recommended solution should be the one that best achieves the goals at the least cost. A plan for paying for the recommended solution is essential to showing that the goals are achievable. The financing plan should address each project within the plan, its schedule for completion, its capital costs and its annual operation and maintenance costs. A strategy for implementing the rehabilitation plan is essential. It should identify who-will do what-by when. It should address all of the elements listed above and be as detailed as necessary to insure the work will get done. The implementation strategy need not be completed when work begins but it should address each element to some degree. It will evolve as work progresses so that at some point in time, it will be clear as to who-will do what and when. The rehabilitation plan should identify measures for determining if the plan has been successful. Have the goals been achieved? Are partnerships flourishing? Has the funding occurred as proposed? The measures should be monitored during the life of the plan and a periodic status report prepared [15] .

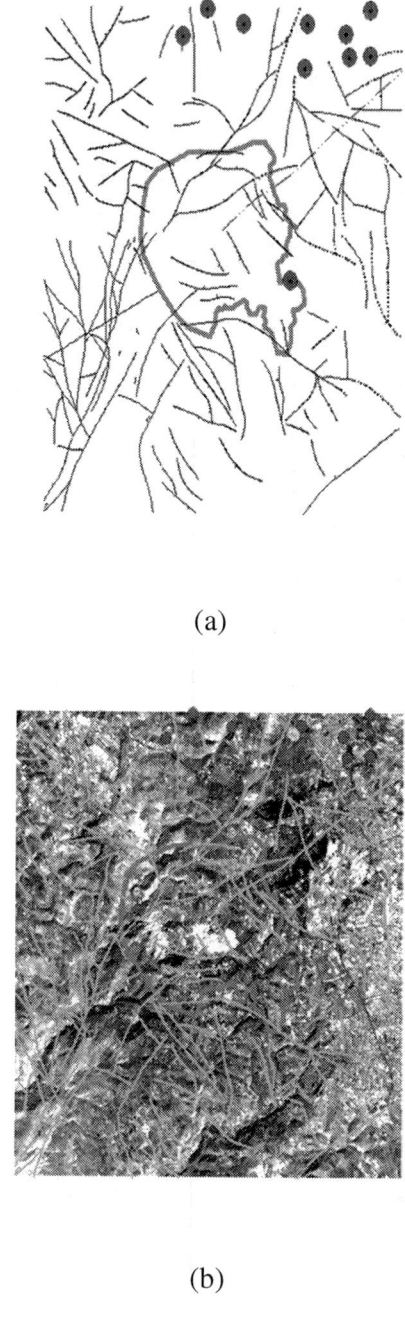

(a)

(b)

Figure 13: Geo-referenced structural map with digitized faults.

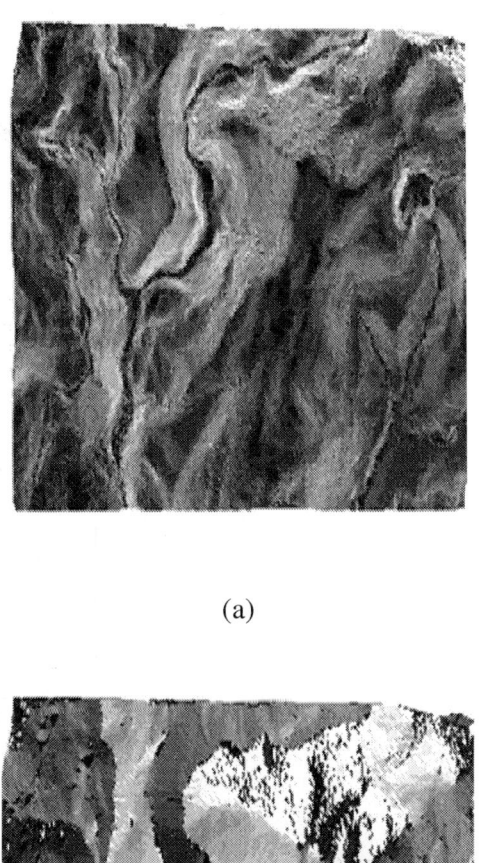

(a)

(b)

Figure 14: (a) Slope map; (b) Aspect map.

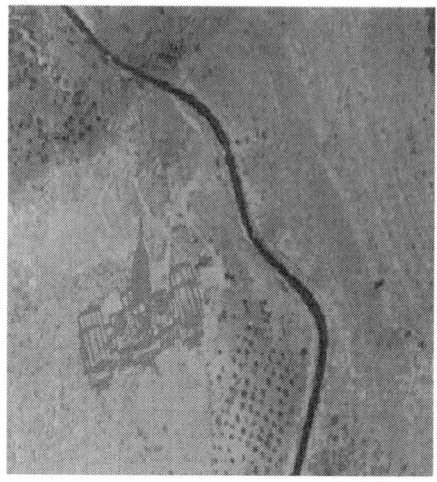

(a)

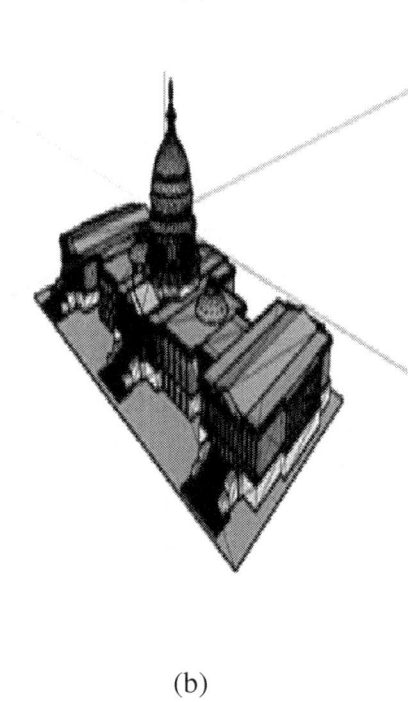

(b)

Figure 15: (a) 3D-model dropped over ortho-photo; (b) rendered 3D-Model after transformation.

Figure 16 shows one of the suggested models that have the following considerations:

- Improve the environment, keep the sense of sustainability local species, and increase the quality of life for population of region.
- Increase the attraction of the region due to natural potentials like vertical cuts which allow new kinds of sports like climbing, mountain biking.
- This module contains school, school Garden and small forest. The aim of this module is to enhance educational diversity by teaching outside the classroom. Through a personal contact with natural potentials the visitors can create a further awareness for ecological issues and an understanding. The people should learn to understand the interrelation between the ecological, economic and social or rather cultural activities.
- This model Avoid wild settlements and to lead developmental necessities in to controllable areas.

CONCLUSION AND FUTURE RECOMMENDATION

As mentioned before the main object was to re-develop the mining area in Mahis. To achieve this Photogrammetry, Remote Sensing and GIS, techniques were applied to the study. The topography of the area and the geological effects in terms of faults and structures have been explored. The geomatics tools facilitate the exploration of the area and assessment procedures that required for rehabilitation process. It has been shown that geomatics techniques were necessary and suitable to establish geodatabase needed by the decision-makers.

It has been shown that the distribution of the ground control points is very important in Photogrammetric Process moreover the quality of the Ortho-photo depends on the quality of the DTM. Future recommendations would focus on building a true Ortho-photo for the mining area in Mahis. Make questionnaire to know exactly what people in the study area need to be executed. Building several suggested modules according to local priorities and the financial opportunities. A developmental concept should consider the transportation issues as part of the needed geodatabase.

Figure 16: One of the suggested model.

ACKNOWLEDGEMENTS

The authors would like to acknowledge the continued support of Al-Balq Applied University and also we would like to acknowledge both Engineer Rana Al-Hadidi and Rawan Hamzah for efforts in data processing throughout this study.

REFERENCES

1.   Wikipedia (2012) Land Rehabilitation. http://en.wikipedia.org/wiki/Land_rehabilitation

2.   Natural Resources Canada (2012) http://www.nrcan.gc.ca/home

3.   Ministry of Northern Development and Mines (2012) http://www.mndm.gov.on.ca/en

4.   Chen, L.C. and Lee, L.H. (1993) Rigourous Generation of Digital Orthophotos from SPOT Images. Photogrammetric Engineering and Remote Sensing, 59, 655-661.

5. Habib, A.F., Ghanma, M.S., Al-Ruzouq, R.I. and Kim, E.M. (2004) 3-D Modelling of Historical Sites Using Low-Cost Digital Cameras. XXth Congress of ISPRS, 12-23 July 2004.

6. Krishna, B.G., Amitabh, T.P., Srinivasan, P. and Srivastava, K. (2008) DEM Generation from High Resolution MultiView Data Product. The International Archives of the Photogrammetry, Remote Sensing and Spatial Information Sciences, 37, 1099-1102.

7. Wolf, P.R. and Dewitt, B.A. (2000) Elements of Photogrammetry with Applications in GIS. 3rd Edition.

8. Hofmann-Wellenhof, B., Lichtenegger, H. and Collins, J. (1997) Global Positioning System Theory and Practice. 4th Edition.

9. Sabins, F.F. (1988) Remote Sensing: Principles and Interpretation. 2nd Edition.

10. Lerma, J.L., Ruiz, L.A. and Buchon, F. (2000) Application of Spectral and Textural Classifications to Recognise Materials and Damages on Historical Building Façades. International Archives of Photogrammetry and Remote Sensing, 33, 480-484.

11. Drury, S.A. (1987) Image Interpretation in Geology. 2nd Edition, London.http://dx.doi.org/10.1007/978-94-010-9393-4

12. Aronoff, S. (1989) Geographic Information System. WDL Publication, Ottaw, 19.

13. Bell, F.G. (1993) Engineering Geology.

14. Wikipedia (2013) Triangulated Irregular Network.http://en.wikipedia.org/wiki/Triangulated_irregular_network

15. El-Hakim, S., Beraldin, A. and Picard, M. (2002) Detailed 3D Reconstruction of Monuments Using Multiple Techniques. ISPRS/CIPA International Workshop on Scanning for Cultural Heritage Recording, Corfu, 58-64.

16. A Rehabilitation Plan for Abandoned Mine Reclamation.http://saint.psend.com/Appendix%20A.htm

Chapter

5

Impact of Underground Mining on Shaft Lining and Aquifer in Eastern China

Qing Yu, Hideki Shimada, Takashi Sasaoka, and Kikuo Matsui

Department of Earth Resources Engineering, Faculty of Engineering, Kyushu University, Fukuoka, Japan

ABSTRACT

Serious shaft lining ruptures have often occurred in the eastern part of China since 1987 due to the complicated geological conditions. This paper tries to find out the relationship between mechanisms of shaft lining rupture and the underground mining process. The analysis is based on the existence typical engineering and geological conditions in eastern China; the impact of underground mining on the shaft lining and aquifer layer is analyzed by using numerical method. The impact factors such as different depths, thicknesses, mining widths of coal seam and different distances to the shaft are used in the analysis. The mining area under the aquifer which near the shaft lining has a significant impact on the shaft lining due to mining process, and increases the risk of occurrence of shaft lining rupture.

INTRODUCTION

Serious shaft lining ruptures have often occurred in the eastern part of China and this problem has been one of the major subjects in Chinese coal mine industry for past twenty years. All these sites that shaft lining rupture occurred are in similar geological conditions. All these shafts pass through deep alluvium of Quaternary strata which the composition of the bottom aquifer is complex [1].

According to the statistical data of the shaft lining rupture in alluvium, the elapsed time after drivaging and location of rupture has its own rules. Generally, the type of shaft lining ruptures is divided into three phases [2]: 1) Rupture during the freezing shaft sinking: horizontal cracks caused by temperature variation is initiated and propagated; 2) Frozen wall melting: horizontal and inclined cracks caused by temperature variation are initiated and propagated; 3) Surface subsidence after several years: horizontal and inclined cracks caused by surface subsidence are propagated, especially, this phenomena is occurred near the boundary between aquifer and bedrock.

The shaft lining rupture happening after several years running is the most common situation in the eastern part of China, such as Datun, Xuzhou, Huaibei, Yanzhou, Yingxia, Hebi, Dongrong, etc., since 1987 [3-5]. When shaft lining rupture occurs, the inner shaft lining delaminates and spalls, longitudinal steel bows inward, transverse cracks form and intersect in the horizontal direction along circle, seepage occurs or even sand gushes, and mostly seriously, concrete blocks fall out and break the equipment in shaft. In addition, the shaft bends up; cage guides, drainpipes and pressure ventilation pipes are in longitudinal bending, and in serious cases the cage is stuck due to torsional deformation.

According to several surveys on coal mines, before the shaft lining rupture happening, the groundwater leakage has been observed in the mining area from roof, the groundwater level decline is happening at the same time in the aquifer [6-9]. The current researches are mostly focused on the causes, mechanisms and solutions for these kinds of geotechnical issues and they have been the major topics in recent 20 years. In this research, the model of strata layers is built and used for numerical simulation, and the impact of coal mining in the deeper coal seam before aquifer drawdown on shaft lining is examined.

ANALYSIS MODEL

FLAC[3D] is a three-dimensional explicit finite-difference program for engineering mechanics computation and simulating the behavior of three-dimensional structures built of soil, rock or other materials that undergo plastic flow when their yield limits are reached.

The basic model used in the analysis is shown in Figures 1 and 2. This model represents the typical engineering and geological conditions in eastern China especially Baodian coal mine which happened the shaft lining rupture in 1995. The net diameter of the main shaft in this mine is 6.5 m and the thickness of the shaft lining is 1.0 m constructed by C28 concrete. The thickness of alluvium is 148.69 m and it contains 3 aquifers. The bottom aquifer is important to the mine and the water level is affected by mining, so in this model, only the bottom aquifer is left for simulation.

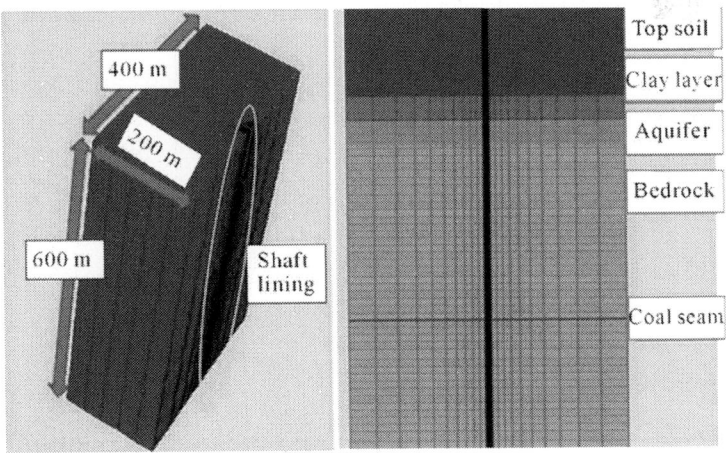

Figure 1: Analysis model.

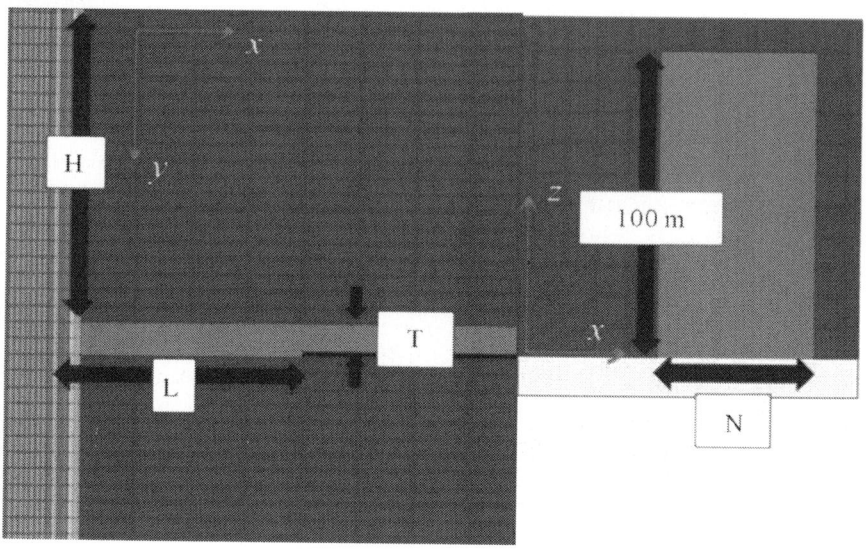

Figure 2: Mining area model.

According to the engineering geological data and shaft construction data of coal mine, the required mechanical parameters for numerical simulation are shown in Table 1.

As the process of soil deformation and shaft lining stress change is very complex, the following assumptions are made:

- Shaft lining, soil and loss areas are treated as the symmetrical distribution, belong to the space axial symmetry.
- Concrete shaft lining, the surrounding soil and aquifer are homogeneous, isotropic. No interface between the surrounding soil and shaft lining, that means no sliding in the interface.

Then the boundary conditions are set as follows: a free top surface, two horizontal fixed slides and a vertical fixed the lower surface.

As the coal seam is under the deep bottom aquifer, the different depth, thickness, mining width of coal seam and different distance to the shaft were simulated as the impact factors of the shaft lining and the aquifer (as shown in Table 2).

ANALYSIS RESULTS AND DISCUSSION

Failure Initiation Due to Mining Process

The shear strain change is chosen to reveal the impact of the mining process, as the rock compression failure mainly belongs to the shear failure [10]. Figure 3 shows the shear strain distribution around the mining area and the aquifer during mining. The high strain area is in yellow color and the low strain area is in blue. From this figure, the both ends of the mining area and the aquifer layer have the high strain. In addition, the shear strain in the rock layer between the mining area and the aquifer exceeds 2.0×10^{-3}. And except the part shown in blue, there is a risk of failure occurs, especially the area between mining area and aquifer. Therefore, the leakage occurs at the impermeability rock layer can be concluded by the fissure developing in the rock which the failure happened during the mining process. According to the above analysis, during the underground mining in 250 m depth, the bedrock under aquifer begins to get loose and then some fissure develops, as a result the aquifer drainage occurs.

Figures 4 and 5 show the distribution of shear strain in different mining depth. The deeper of the mining area, the distance to the aquifer is greater, then the impact on the aquifer is gradually reduced. The impact on the aquifer becomes disappear while mining in 400 m depth, and the shear strain near the mining area tends to become higher. For this reason, if the mining depth is deeper than 400 m, it becomes much more important to consider the safety of the surrounding area compare with the impact on the aquifer.

Table 1: Mechanical parameters of soil and shaft lining

Material	Top soil	Clay	Aquifer	Bedrock	Shaft lining	Coal seam
Thickness (m)	120	30	30	450	1	
Dry density (g/cm³)	1.65	1.65	1.7	2.7	3.0	2.7

Bulk modulus (Mpa)	50	50	61.25	1.1×10^4	$2 < 10^4$	1.1×10^4
Shear modulus (Mpa)	23	23	28	8333	12600	8333
Porosity	0.18	0.05	0.3	-		-

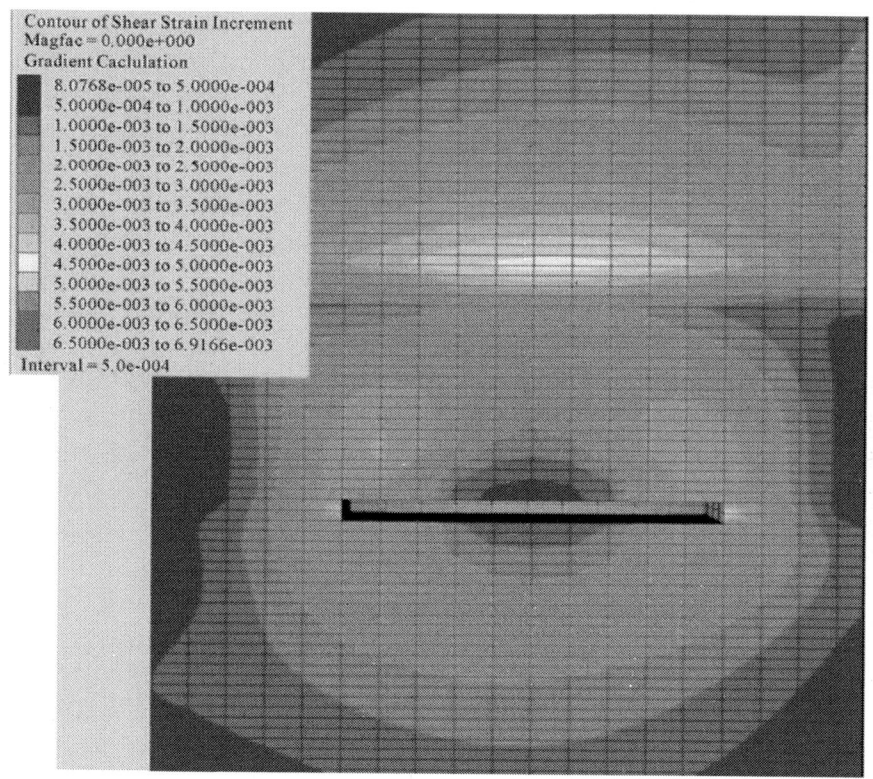

Figure 3: Shear strain distribution during mining (50 m to the shaft and 250 m mining depth, 6 m thickness coal seam).

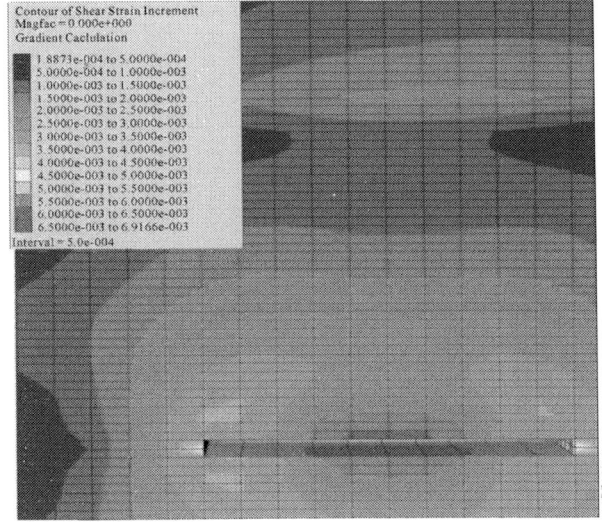

Figure 4: Shear strain distribution during mining (50 m to the shaft and 300 m mining depth, 6 m thickness coal seam).

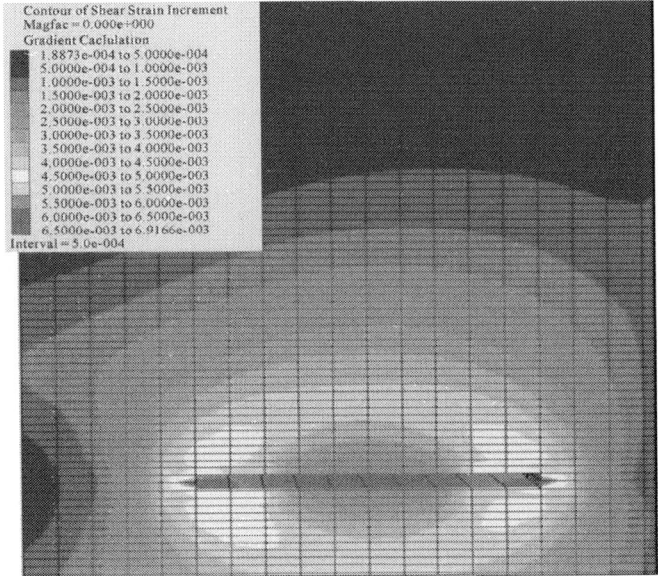

Figure 5: Shear strain distribution during mining (50 m to the shaft and 400 m mining depth, 6 m thickness coal seam).

Table 2: Factors for analysis

Depth of coal seam (H)	250 m × 300 m × 400 m
Thickness of coal seam (T)	3 m × 6 m × 9 m
Mining width (N)	50 m × 100 m
Distance to shaft (L)	50 m × 70 m × 90 m

Impact Due to Mining Distance to Shaft

The stress in shaft lining is the most concerned issue, so the impact to the shaft lining is revealed by stress change. Figure 6 shows the impact on the surrounding areas and the shaft lining by different distance of mining area to the shaft. The lines in the figure are the values of maximum principal stress in the surrounding area, aquifer and shaft lining. As the distance between mining area and shaft becomes greater, the maximum principal stress in shaft lining becomes smaller, the mining impact on the shaft lining becomes smaller and that can reduce the concentration of the maximum stress.

Impact Due to Mining Depth

The mining depth also has the impact to the stress in the shaft lining, though Figure 7 shows the maximum principal stress change due to different mining depths. As the mining area is deeper and the overburden increases, the deformation of surrounding area is greater and the stress concentration in mining area and shaft lining tends to rise as the depth increases. Also the maximum principal stress in the mining area is higher than in the shaft lining. From the above analysis, the risk of shaft lining rupture is smaller than the risk of mining area collapse. So the coal pillar and artificial support need to be remained in the mining area considering the collapse risks.

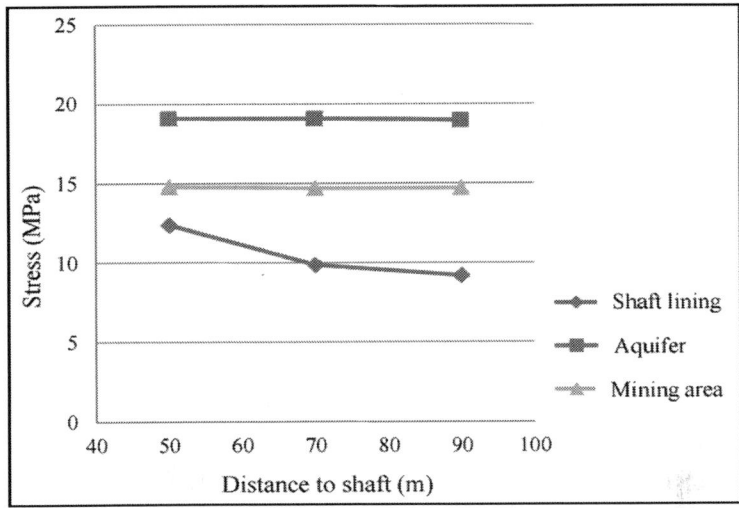

Figure 6: Impact due to distance to shaft (250 m mining depth).

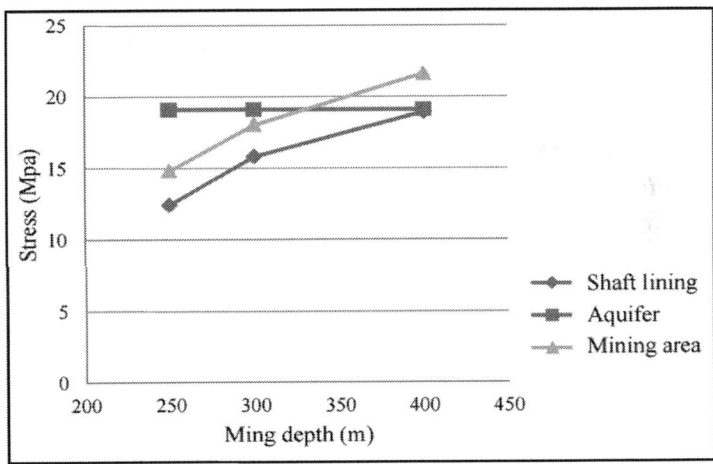

Figure 7: Impact due to mining depth (50 m to shaft).

Impact Due to Mining Width

The change of mining width leads to the deformation of the ground layers; the displacement of the ground layers is used to show the impact

Figures 8 and 9 show the influence of 50 m and 100 m mining width As the mining width increases, the deformation of surrounding mining area increases which it also gives an impact on the deformation of the shaft lining. As shown in Figure 10, when the mining width is 50 m, the maximum principal stress concentration occurs in the surrounding area, and when the mining width is 100 m, the maximum stress occurs in the shaft lining.

Therefore, due to the deformation of the mining area, there is a significant stress concentration occurs in shaft lining. Particularly, a great stress concentration has occurred around the mining area and the impact range of 50 m mining width and 100 m mining width has a large difference. From above analysis, the mining width has a significant impact on the shaft lining, and the failure in shaft lining is much more severe than in the mining area.

Impact Due to Coal Seam Thickness

Three thicknesses of coal seam are set as 3 m, 6 m and 9 m for examining the impact. And the stress change in the shaft lining, aquifer and mining area is used to reveal the impact of the coal seam thickness.

Figure 11 shows the maximum principal stress in the mining area, aquifer and shaft lining. From the figure, the stress in aquifer has no obvious change. However, the stress in mining area has a significant increase with the increase of the coal seam thickness. Furthermore, the stress in mining area has outstripped the stress in shaft lining. That means the thickness change of coal seam has an impact on the mining area, but not obvious in the shaft lining and aquifer.

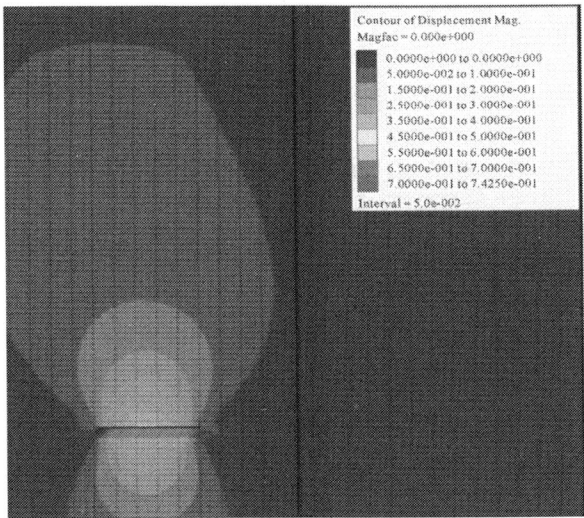

Figure 8: Vertical displacement distributions under 50 m mining width (50 m to the shaft and 400 m mining depth, 6 m thickness coal seam).

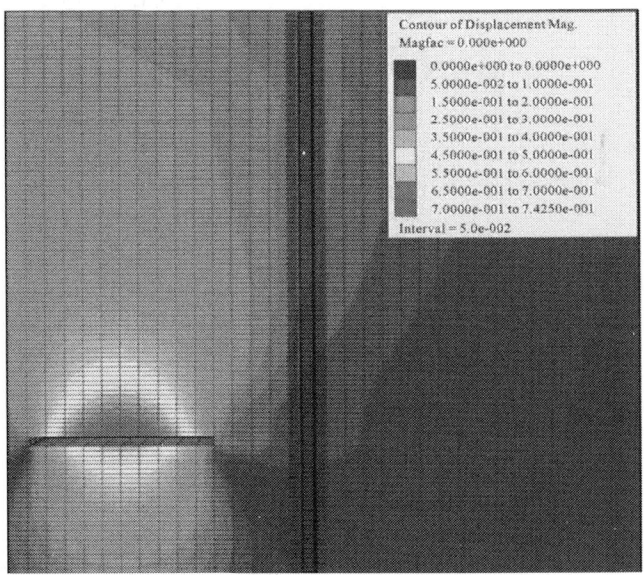

Figure 9: Vertical displacement distributions under 100 m mining width (50 m to the shaft and 400 m mining depth, 6 m thickness coal seam).

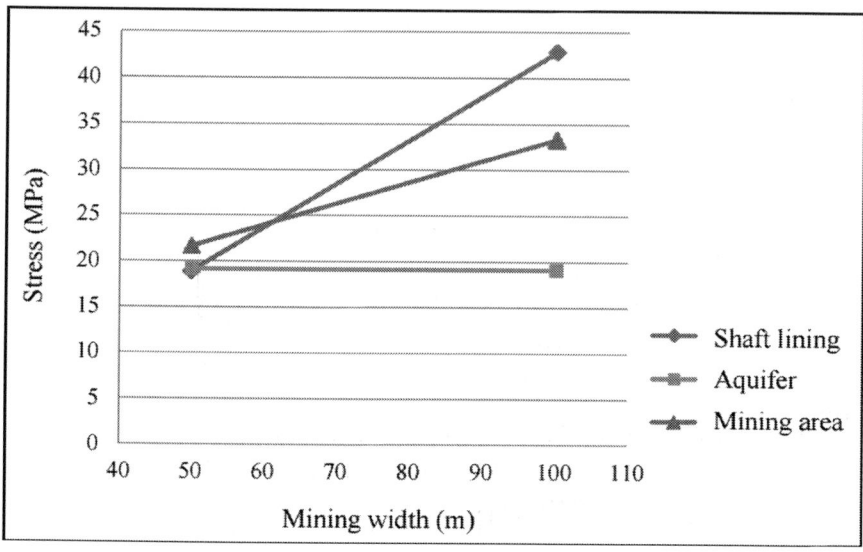

Figure 10: Impact due to mining depth due to mining width (50 m to the shaft and 400 m mining depth, 6 m thickness coal seam).

CONCLUSIONS

In this study, the impact of the underground mining on the shaft lining and aquifer is analyzed by numerical simulation. The main aim is to find out the relationship between mechanisms of shaft lining rupture and the underground mining process. The impact factors such as different depth, thickness, mining width of coal seam and different distance to the shaft are used in the analysis. From a series of numerical simulations, the impact of different factors can be cleared as follows:

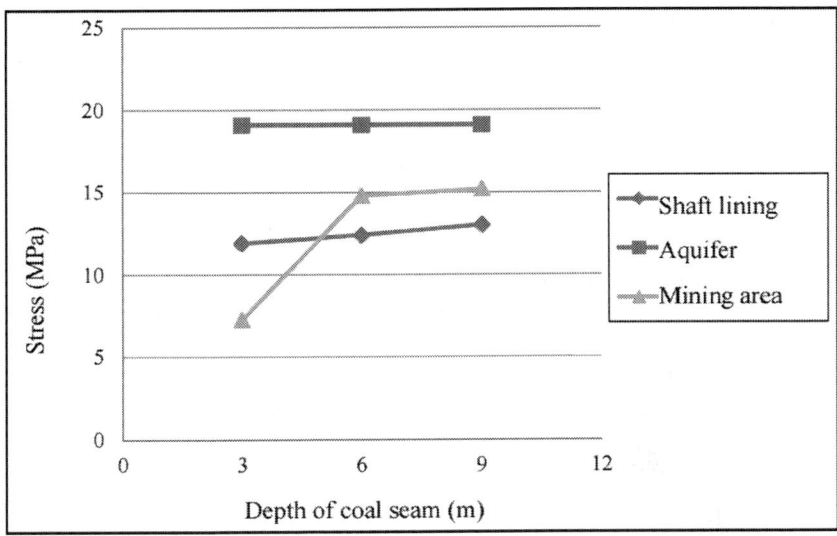

Figure 11: Impact due to coal seam thickness (50 m to shaft and 250 m mining depth).

- The underground mining process can lead to the initiation of the failure in the soil and rock layers. If the mining location is near some certain range of the aquifer, the aquifer drainage will occur due to the fissure in aquifer connection to fissures in the bedrock. This situation should be avoided in the mining process by changing the mining location or using supporting structure.
- The impact of the mining distance to the shaft lining is obvious, if the mining location is closed to the shaft lining, the greater stress concentration will happen in shaft lining and increase the risk of occurrence of shaft lining rupture.
- When the mining location is near the shaft ling in a certain range, the increase of the mining depth also can increase the risk of occurrence of shaft lining rupture. More seriously, the possibility of mining area collapse is raised due to the increase of mining depth.
- The mining width has a significant impact on the shaft lining, the stress in shaft lining increase fast due to the increase of mining width. Before the increase of the mining width, the support structure should be prepared to protect the mining area in case the stress transfers to shaft lining.

- The thickness change of coal seam has an impact on the mining area, but not obvious in the shaft lining and aquifer.

Therefore, when the mining area under the aquifer nears the shaft lining, there is a significant impact on the shaft lining due to mining process, and the risk of occurrence of shaft lining rupture become higher and need to be concerned.

ACKNOWLEDGEMENTS

Authors are grateful for financial assistance provided by the Global-Centre of Excellence in Novel Carbon Resource Science, Kyushu University.

REFERENCES

1. G. Cui and X. Cheng, "Occasions of Damaging Shaft Walls in Xuhuai District," Coal Science and Technology, Vol. 19, No. 8, 1991, pp. 46-50.

2. L. Jing and H. Wang, "The Research about the Fracture Regularity of the Frozen Shaft Wall in Surface Soil," Journal of Huainan Institute of Technology, Vol. 20, No. 1, 2000, pp. 14-20.

3. M. Gu, "Failure of Shaft Caused by Special Geologic Hazards and Corresponding Prevention Measures," Journal of Engineering Geology, Vol. 8, No. 2, 2000, pp. 197-201.

4. W. Yang and H. Fu, "Theoretical Investigation on Vertical Additional Force on Shaft Lining in Special Stratum," Journal of China University of Mining & Technology, Vol. 9, No. 2, 1999, pp. 129-135.

5. G. Zhou and G. Cui, "Numerical Analysis on Interaction between Shaft Wall and Surrounding Strata after Aquifer Grouting," Journal of China University of Mining & Technology, Vol. 27, No. 2, 1998, pp. 135-139.

6. H. Liu, W. Chen and Z. Wang, "Theoretical Analysis of Shaft Lining Damage Mechanism of Yanzhou Mine," Chinese Journal of Rock Mechanics and Engineering, Vol. 26, Supp. 1, 2007, pp. 2620-2626.

7. S. Bi, X. Lou and B. Xu, "On the Mechanism of Coal Mine Shaft Damage Caused by Subsidence in Xuhuai Area, Southeast China," Communications in Nonlinear Science & Numerical Simulation, Vol. 2, No. 2, 1997, pp. 75-80. doi:10.1016/S1007-5704(97)90043-5

8. Yang W., Cui G. and Zhou G., "Fracture Mechanism of Shaft Lining under Special Strata Condition and the Technique Preventing the Shaft From Fracturing (Part One)," Journal of China University of Mining & Technology, Vol. 25, No. 4, 1997, pp. 1-4.

9. G. Zhou and X. Cheng, "Study on the Stress Calculation of Shaft Lining Surrounded by Special Strata," Journal of China University of Mining &Technology, Vol. 24, No. 4, 1995, pp. 24-30.

10. M. You and A. Hua, "Fracture of Rock Specimen and Decrement of Bearing Capacity in Uniaxial Compression," Chinese Journal of Rock Mechanics and Engineering, Vol. 17, No. 3, 1998, pp. 292-296.

Environmental Impact on Surface and Ground Water Pollution from Mining Activities in Ikpeshi, Edo State, Nigeria

G. N. Idris[1], G. O. Asuen[2], and O. J. Ogundele[1]

[1]Department of Minerals and Petroleum Resources Engineering, Auchi Polytechnic, Auchi, Nigeria
[2]Department of Geology, University of Benin, Benin City, Nigeria

ABSTRACT

The study was carried out to evaluate the surface and groundwater condition from mining activities in Ikpeshi and its environs in Akoko Edo Local Government Area of Edo State, Nigeria. Twenty water samples were randomly collected and analyzed—one borehole water sample, two hands dug wells, eight river samples and nine quarry pits water samples. The physiochemical, heavy metal and bacteriological analysis of the water sample, as well as the variables were compared with those of the World Health Organization (WHO) standard (2008), United State Environmental Protection Agencies (USEPA) standard (2012) and National Agency For Food, Drug Administration And Control (NAFDAC) in Nigeria to determine their suitability for drinking and domestic purposes. The variables determined are: pH ranges from 7.67 - 8.56 mg/l which is suggestive of neutral to alkaline in character, calcium ranges from 5.12 - 2416 mg/l, turbidity ranges from 1.16 - 15.32 mg/l, total dissolved solid (Tds) ranges from 90 - 366 mg/l and total hardness ranges from 58.65 - 187.37 mg/l, fall within WHO standard, are suggestive of concentration of detergent from soap, calcium, magnesium, suspended solid particles and colloidal matters from some of the water samples. While iron ranges from 0.08 - 0.16 mg/l, potassium ranges from 0.02 - 0.18 mg/l, chloride ranges from 30.03 - 120.13 mg/l, sulphate ranges from 1.03 - 5.36 mg/l, nitrate ranges from 0.01 - 0.23 mg/l, lead ranges from 0 - 0.01 mg/l, Zinc ranges from 0 - 0.08 mg/l, copper ranges from 0 - 0.02 mg/l and magnesium ranges from 1.38 - 6.56 mg/l, fall within standards. Coliform count ranges from 0 - 14 mg/l. The water should be treated before the consumption because of its high concentration of detergent, suspended particles, faecal materials and calcium from the water samples. The quarry pits should be reclaimed and rehabilitate after mining. Alkaline materials should be used to neutralize the rock pile area, dumped site, tailing and mine pit itself to avoid acid generation.

INTRODUCTION

In any environment, there is a strong relationship between human activities and water pollution of that environment due to anthropogenic activities resulting from the growth of industries and technological advancement.

The sources of pollution in mining terrain in Ikpeshi are open pits, waste disposal area, haulage roads, processing plant mills, tailing, and waste rock piles area.

Direct degradation can occur to ground water situation downhill from a surface mine by the flow of contaminated drainage from the mine. This mine drainage can come from pits, ponds or from rain fall infiltration and groundwater flow during mining and often reclamation. Groundwater pollution would result from the same toxic overburden.

Indirect degradation of groundwater could result from blasting which causes a temporary shaking of the rock and results in the new rock fracture near working area of the mine. Blasting can also cause the old pre-existing rock fracture to become more open or permeable, by loosening mineral debris or cement in this fracture; this could affect nearly vertical fractures located up to several hundred feet away from the surface mine causing vertical leakage of pond mine drainage from nearby abandoned deep mines to underlying aquifers.

Water is considered polluted when it is altered in composition or condition direct or indirectly as a result of activities of man so that it becomes unsuitable or less suitable for any or all of the functions or purposes for which it would be suitable in its natural state [1] . Pollution of water whatever description is actionable [2]. Environmental pollution and contamination are becoming a common occurrence in part of developing ward [3]. Chemistry of ground water, its classification, standard and usage studies were carried out [4] , undertook a case study of Nigeria. In tropical countries like ours, the high temperature partially accounts for the water dissolving more materials. Elements like lead, copper, cadmium, chromium etc. are regarded as toxic elements. Therefore, their presence in water above permissible limits stipulate by WHO and NAFDAC which are peculiar to Nigeria can affect the central nervous system [5].

Ikpeshi mining companies are the largest in Edo State, Nigeria and have a processing capacity of 175,000 tons per day of crushed ore. The catchment area contains mineralized rocks which contain solid minerals, and usually have elevated metal level. As the trace metal content of river water is normally controlled by the abundance of metal in the rock of the catchment area and by their mobility.

The land surface of the study area is dominated by sand stone ridges to south, and there are generally undulating schist outcrop around Ikpeshi town. Marble is located in the north-eastern part of the area and this is where the quarry operation is mainly located. Most of the stream is fed by spring water and see pages through some joints and fracture zones in the underlying basement rocks.

METHODOLOGY

Reconnaissance survey was carried out to locate strategic position where samples can be collected so as to obtain reliable result.

Polythene bottles containers were used for the collection of water samples at each location, samples were taken at the rivers, bore hole, hand dug wells and open quarry sites. Before collection of samples, the polythene bottles containers were rinsed again by the water from which samples were to be taken. This was done for each location where samples were collected. Samples were collected at different locations randomly.

The samples were duplicated and corked tightly and well labelled according to their location and number. The samples were kept in refrigerator at a temperature of about 4°C to prevent further bacterial replication before they were taken to the laboratory for further analysis on the next day.

The refrigerator of the water is a means of preventing the spontaneous increase of bacterial species in the water and to maintain the natural state of the water samples.

Determination of pH in Water (Apha 460)

To measure the pH of the samples collected, pH meter (Model Testr-1), Beaker (100 ml capacity) and Buffer solutions of pH 4.01 and 10.01 were used. At a given temperature, the intensity of the acidic or basic character of a solution is indicated by hydrogen ion activity. In the field and laboratory, pH is electrometrically measured using a pH meter with a glass electrode. pH meter was calibrated and the following procedures were followed to measure water samples pH level:

Ÿ The pH meter electrode was used copiously rinsed with distilled water.

Ÿ Pour about 100 ml of water samples into a clean 100 ml capacity beaker.

Ÿ The electrode end of the pH meter was inserted into the sample and pH reading was taken after wards.

Ÿ Both the pH meter electrode and the samples were copiously rinsed with distilled water.

Ÿ 100 ml of water samples was poured into a clean 100 ml capacity beaker.

Ÿ The electrode end of the pH meter was inserted into the sample, pH readings taken.

Ÿ All the pH values were recorded and tabulated.

Determination of Electrical Conductivity in Water

Electrical conductivity of water is a numerical expression of the ability of an aqueous solution to carry an electric current. To determine electrical conductivity in water, electrical conductivity meter (pioneer 65) and 0.01 M KCl solution were used. The equipment was calibrated. The following procedures were followed to measure the samples:

Ÿ Rinse the EC meter electrode copiously with distilled water, followed by the sample.

Ÿ Pour about 100 ml of water samples into a clean 100 ml capacity beaker.

Ÿ Insert the electrode end of the meter into the sample, press READ button and wait for a stable reading.

Ÿ Record the EC value.

Lovibond comparator and turbidity meter model 800 were used to determined colour measurement and turbidity respectively. Also, total hardness was determined using buffer solution, 2-drops of eriochrome black T indicator was added and titrated with 0.01 N EDTA from wine colour to blue end-point.

Determination of Salinity as Chloride Water

Mohr method was employed which involved silver nitrate as titrant and potassium chromate as the end point indicator. The chloride ion present in the wastewater sample is precipitated as white silver chloride.

Colorimetric method was employed in the determination of sulphate ion. Volumetric flasks (100 ml capacity), pipettes (10 ml capacity), cuvette (25 ml capacity) and HACH DR2000 (absorbance mode) were used with the following reagents: barium chloride salt, NaCl-HCl solution and alcohol-glycerol mixture.

Determination of Phosphate in Water

HACH DR2000 (Absorbance Mode), 50 ml Volumetric flasks, 10 ml pipette, 25 ml cuvette, Hot plate and 250 ml round bottom flask were used with the following reagents/chemical: ammonium molybdate-antimonyl solution, Ascorbic acid (2% w/v) solution, Phosphate standard stock solution (1000 mg/L), perchloric acid, NaOH solution, 6 M and methyl orange indicator.

Oxidation method, multiple tube test (APHA 9222A) and brucine method were used to analysis heavy metals in water wet, total coliform bacteria and nitrate in water.

RESULTS AND DISCUSSION

The obtained data from water analysed are showed in Tables 1-3.

The result shows that turbidity, hardness, TDS, magnesium and calcium are very high in some samples. That calcium had the greatest values both for fresh water and drinking water respectively. That the environment is more of calcite than dolomite make it calcitic dolomite environment. Copper, Zinc, pH, conductivity, hardness, chloride, and sulphate occur in reasonable concentrations which are within the average values as compared to WHO, NAFDAC and USEPA values. From the potential health effects like urinary tract infection, bacteraemia and diarrhoea from the injection of water containing such concentration above mentioned parameters, the water collected should be treated before consumption.

The colour of the samples collected, ranges from 0 - 1 hazen units were not all cleared as a result of high concentration of dissolved calcium and magnesium including sample SP4 which is the borehole. The pH of surface and groundwater in the study area generally ranges from 7.61 - 8.65, which means that the pH is generally neutral to alkaline. Electrical conductivity generally ranges from 182.1 - 733 (μS/m), which means the sample are good conductor of electricity (Table 4)

The result of the water analysis in the study area show that pH of the groundwater and surface water ranges from 77.56 - 187.37, meaning that the water hardness falls within moderately hard water, hard water and very hard water. The Iron in the sample ranges from 0.08 - 0.16.

Table 1: Laboratory data for samples analysed

Sample code	Ph	Temp (°C)	TDS (mg/l)	Cond (µs/cm)	Colour (Lu)	Turbidity (N.T.U)	Total hardness (mg/l)	Ca (mg/l)	Mg (mg/l)	Na (mg/l)	K (mg/l)	Chloride (mg/l)	TOC (mg/l)	Sulphate (mg/l)	Phosphate (mg/l)	Nitrate (mg/l)
1-SWS	7.91	29.4	247	496	1	15.32	109.72	17.61	5.92	7.21	0.18	85.05	0.01	2.43	0.51	0.09
2-RWS	8.32	29.1	183	336.2	1	2.31	117.89	16.64	4.98	5.42	0.15	80.02	0.03	1.72	0.18	0.03
3-QWS	8.02	28.6	249	498	0	1.43	178.32	20.03	7.62	7.71	0.16	90.08	0.02	1.57	0.14	0.07
4-HDW	7.67	28.9	366	733	1	13.52	187.37	24.16	6.56	8.35	0.11	120.13	0.07	5.36	1.63	0.23
5-BHW	7.81	29.6	90	182.1	0	1.72	58.65	5.12	1.58	2.18	0.09	30.03	0.01	2.15	0.93	0.01
6-QWS	7.78	30.4	129	258.5	1	8.64	104.25	10.88	3.34	4.21	0.10	65.11	0.02	1.03	0.07	0.03
7-QWS	8.22	29.6	178	354.2	0	1.57	115.97	14.45	4.63	4.28	0.13	75.26	0.01	1.17	0.11	0.02
8-QWS	8.56	28.1	113	224.9	0	1.38	77.56	7.23	1.93	2.81	0.05	57.94	0.01	1.36	0.13	0.01
9-QWS	8.45	28.4	114	228.3	0	1.16	84.53	7.52	2.23	1.72	0.03	61.25	0.03	1.04	0.07	0.01
10-RWS	8.12	28.2	227	453	1	4.16	127.36	22.83	1.38	2.39	0.02	78.14	0.02	1.24	0.03	0.02
11-SWS	8.41	28.3	251	421	1	1.63	171.89	15.21	4.23	6.35	0.17	80.13	0.02	1.33	0.16	0.04
12-SWS	8.56	28.4	117	231.6	0	1.41	53.87	13.42	1.45	5.14	0.04	52.12	0.01	1.18	0.08	0.03
13-QWS	8.01	29.3	216	412	1	1.91	108.93	8.32	1.83	4.36	0.16	71.13	0.03	1.06	0.12	0.01
14-HDW	7.72	28.6	296	728	1	12.24	172.42	18.54	1.32	2.16	0.03	29.58	0.02	2.09	0.92	0.06
15-SWS	8.21	29.4	182	359.2	0	2.21	112.45	21.05	2.12	1.83	0.08	58.86	0.01	1.13	0.14	0.02
16-QWS	8.62	28.2	172	298.4	1	1.52	138.32	6.13	3.15	4.27	0.14	81.56	0.01	1.09	0.02	0.05

17-QWS	7.95	29.2	106	2 4.8	0	14.82	56.47	7.01	4.06	7.43	0.02	92.45	0.03	1.27	0.52	0.01
18-QWS	8.47	28.6	224	443	1	2.41	123.24	19.26	1.98	6.08	0.15	65.34	0.02	1.31	0.15	0.03
19-SWS	8.08	28.9	234	478	1	13.42	162.21	20.06	4.06	7.12	0.13	48.67	0.02	1.05	0.17	0.04
20-RWS	8.01	29.3	257	499	1	8.73	105.47	5.93	1.81	2.13	0.05	76.07	0.01	1.23	0.23	0.07

Table 2: Laboratory result

Sample code	Bicarbonate (mg/l)	Fe (mg/l)	Pb (mg/l)	Mn (mg/l)	Zn (mg/l)	Cu (mg/l)	Coliform count Cfu/ml	Cd (mg/l)	Cr (mg/l)
1-SWS	50.12	0.08	BDL	0.05	0.02	BDL	14	BDL	BDL
2-RWS	40.73	0.12	0.01	BDL	0.06	BDL	11	BDL	BDL
3-QWS	57.45	0.10	BDL	0.02	0.05	0.01	2	BDL	BDL
4-HDW	84.93	0.13	0.01	0.03	BDL	BDL	9	BDL	BDL
5-BHW	24.65	0.09	BDL	BDL	0.03	0.01	4	BDL	BDL
6-QWS	31.83	0.12	BDL	0.02	0.07	BDL	2	BDL	BDL
7-QWS	35.63	0.16	0.01	BDL	0.05	BDL	0	BDL	BDL
8-QWS	21.83	0.09	BDL	0.03	0.03	0.01	0	BDL	BDL
9-QWS	22.25	0.12	BDL	0.03	0.08	BDL	4	BDL	BDL
10-RWS	54.5	0.08	BDL	0.01	0.05	0.02	6	BDL	BDL
11-SWS	38.23	0.08	0.01	0.02	0.03	BDL	13	BDL	0.01
12-RWS	22.09	0.09	0.01	0.04	0.02	BDL	2	BDL	BDL
13-QWS	32.72	0.11	BDL	0.01	0.04	0.01	12	BDL	BDL
14-HDW	55.43	0.09	BDL	BDL	0.03	BDL	4	BDL	BDL
15-SWS	41.86	0.13	BDL	0.03	0.06	0.01	0	BDL	BDL

16-QWS	28.19	0.08	0.01	BDL	0.08	0.02	0	BDL	BDL
17-QWS	50.09	0.12	0.01	0.01	0.03	BDL	2	BDL	BDL
18-QWS	34.28	0.15	BDL	0.05	0.02	0.01	14	BDL	BDL
19-SWS	48.62	0.14	0.01	0.01	0.04	0.02	6	BDL	BDL
20-RWS	51.53	0.09	BDL	0.03	0.05	BDL	0	BDL	BDL

Note: BDL (below detection limit i.e. < 0.001).

Table 3: Comparison of the obtained values with the World Health Organization standard [6]

Parameter	Highest desirable (mg/l)	Maximum permissible level (mg/l)	Range of values obtained from analysis (mg/l)	Remarks
Ph	7.0 8.9	6.5 - 9.5	7.61 - 8.65	It is neutral to alkaline
Conductivity	900 (µS/m)	1200 (µS/m)	182.1 - 733	
Total hardness	100	100	77.56 - 187.37	Moderately to very hard water
Magnesium	20	20	1.38 - 6.56.	Extremely low
Calcium	0.01	0.07	5.12 - 25.16	Highly concentrated
Sulphate	250	500	1.03 - 2.43	Suitable for drinking
Iron	1.0	3.0	0.08 - 0.16	Ok
Nitrate	10	50	0.01 - 0.23	Ok
Lead	0.01	0.01	BDL - 0.01	OK
Cadmium	0.003	0.003	BDL	OK

Chromium	0.05	0.05	BDL	OK
Turbidity	5.0	5.0	1.16 - 15.32	7 samples out of 20 are not suitable
Sodium	-	200	1.72 - 8.35	Suitable
Manganese	0.0	0.4	BDL-0.05	Suitable
Zinc	0.01	3.0	BDL-0.08	Suitable
Copper	0.5	2.0	BDL-0.02	Suitable
Total coliforms	Not allowed	Not allowed	0 - 14	Only 5 samples are ok
Total organic carbon*	-	5	0.01 - 0.07	Ok
Total dissolved solids (mg/l)*		500	90 - 366.	Ok
Potassium**	10.0	11.0	0.02 - 0.18.	Ok

*United State Environmental Protection Agency, 2012. [7]. **National Agency for Food, Drug Administration and Control [8]. BDL: below detection limit.

Table 4: Water classification according to its hardness [9]

Hardness	Types of hardness
0 - 60	Soft water
61 - 120	Moderately hard water
121 - 180	Hard water
>180	Very hard water

The use of fertilizer for agriculture shows little trace of potassium concentration in the study area which ranges from 0.02 - 0.18. The magnesium concentration of the analyzed water sample ranges from 1.38 - 6.56. The concentration of calcium ranges from 5.12 - 25.16 mg/l.

The presence of coliform count bacterial in a water body indicates that the water has contaminated with the faecal material of man or other animals. The presence of faecal contamination is an indicator that a potentials health risk exist for individuals exposed to this water. The values of coliform count, ranges from 0 - 14. No faecal materials are allowed in drinking water but are prominent in some of the samples especially in stream water samples. It presence makes the samples examined contaminated and inadequate for drinking.

The total dissolved solid ranges from 90 - 366 $mg \cdot l^{-1}$ and it falls within the WHO permissible limit of 500 $mg \cdot l^{-1}$.

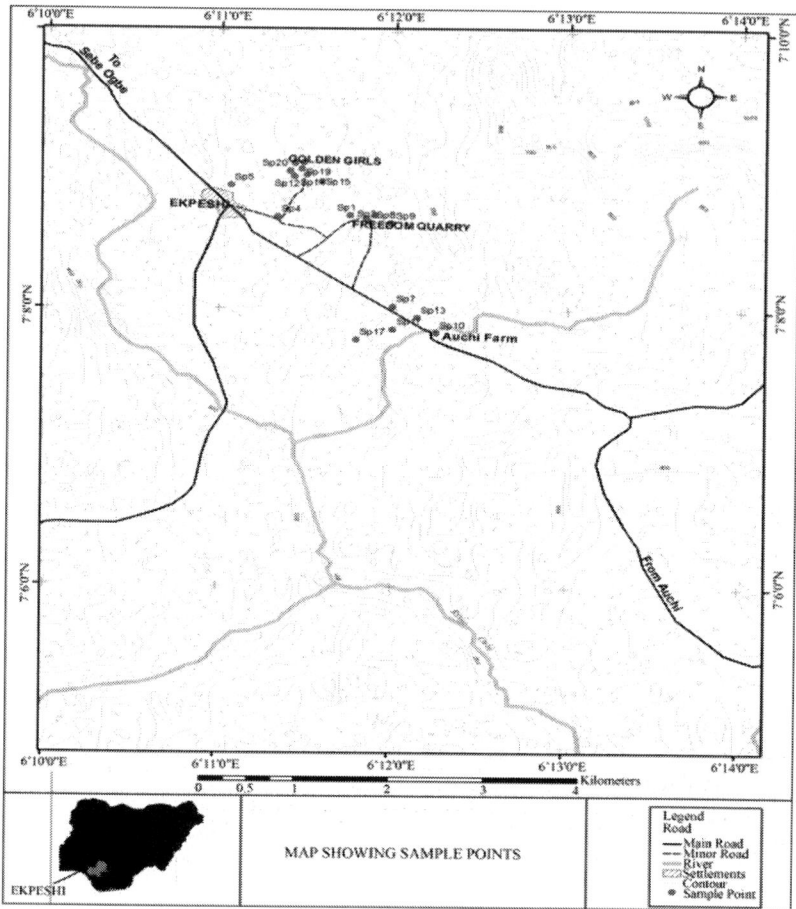

MAP SHOWING SAMPLE POINTS

EKPESHI

Legend
Road
— Main Road
-- Minor Road
— River
Settlements
Contour
• Sample Point

CONCLUSIONS

The findings of this study serve as the basis for making the following conclusions:

- The measured pollutants found their way to the ground and surface water system through path ways such as fracture zones and pore space into the aquifer zone, through run-off, infiltration and percolations thereby contaminating the portable water.
- Turbidity, hardness, total dissolved solids, magnesium and calcium are very high in concentration in some of the samples.

Total hardness was also very high and calcium had the greatest concentration both for fresh water and drinking water respectively.

- Copper, Zinc, pH, conductivity, hardness, chloride, and sulphate occur in reasonable concentrations which are within the average values as compared to WHO, NAFDAC and USEPA standards.
- Faecal materials are higher in the obtained samples that were closer to the settlement than far away from the settlement.

RECOMMENDATIONS

Based on the findings of this study, the following recommendations have been proffered:

- From the potential health effects from the injection of water containing such concentration above mentioned parameters, the water collected should be treated before consumption.
- Total pollution or degradation prevention cannot be attained but the important thing is to reduce the effect of pollution and degradation to the nearest minimum.
- Mine pits should be reclaimed for future use.
- The Environmental Protection Agency of Edo State (EDSEPA) should live up to its statutory obligation of maintaining standards.

REFERENCES

1. Panamello, R. and Mohano, O. (1973) Principle and Problems of Groundwater Resource with Cases Example from Developing Countries. Riparian Right and Pollution.

2. Ezenabor, B.O. (1991) Pollution of Water Whatever Description Is Actionable Riparian Right.

3. Egboka, B.C.E., Nwankwo, G.I., Orajaka, I.P. and Jiofor, A.O. (1989) Guidelines for Drinking Water Quality (World Water) Illinois Groundwater Pollution.

4. Offodile, M.E. (1992) An Approach to Groundwater Study and Development in Nigeria. Mecon Geology and Engineering Services Ltd., Jos.

5. Oteze, G.E. (1981) Water Resources in Nigeria Environment. Vol. 3, Springer, Bertlin Heidelberg, 177-184.

6. World Health Organisation (1994) International Standard for Drinking Water. 8th Edition, Geneva, 53 p.

7. United State Environmental Protection Agency (2012) Edition of the Drinking Water Standards and Health Advisories.

8. National Agency for food and Drugs Administration and Control, Nigeria (2008) To Safeguard the Public Health of the Nation.

9. Hem, J.D. (1970) Groundwater Hydrology. McGraw-Hill, Kogakusha, LTD, Tokyo, 480 p.

Chapter 7

Lateral Stress Concentration in Localized Interlayer Rock Stratum and the Impact on Deep Multi-Seam Coal Mining

Rami Al-Ruzouq[1] and Samih Al Rawashdeh[2]

Mingwei Zhang, Hideki Shimada, Takashi Sasaoka, and Kikuo Matsui
Department of Earth Resources Engineering, Faulty of Engineering, Kyushu University, Fukuoka, Japan

ABSTRACT

To explore the impact of lateral stress concentration in interlayer rock stratum on the exploitation of protected coal seam, a field experiment was carried out in a multi-seam mining structure. Lateral stress redistribution and interlayer rock failure behavior were surveyed. Then an assistant numerical investigation was implemented to evolve the effect of liberated seam mining and its influence on stress reconstruction in surrounding rock mass. The cause of lateral stress concentration and its impact were discussed finally. Key findings turn out that a certain lateral stress increases in interlayer rock stratum and concentrates on its lower region. Lateral stress concentration and interlayer rock failure are interactional. The former is an inducing factor of the latter; the latter promotes the increase of concentration degree. Extent of lateral stress concentration increases to the maximum as seam distance is about 50 m. But the efficacy of liberated seam mining decreases as the seam spacing gets larger. Protected seam mining is then classified based upon the impact of lateral stress concentration, which helps to prevent the rock burst hazard and then to achieve a reliable mining in deep mines.

INTRODUCTION

Many geological hazards distinguished from the conventional ones occur gradually in the underground coal mines with the increase of mining depth [1-3], for instance, the high temperature, high ground pressure, seismic hazard, and rock burst. Among them, rock burst hazard greatly hampers the aim of safe production [4, 5]. How to effectively prevent and control rock burst is becoming a challenging subject for the mining researchers worldwide [6, 7]. In recent years, multi-seam coal mining is treated as an effectively precautionary approach for the deep geologic hazard prevention, especially, for the rock burst [5, 8, 9]. It changes the initial distribution of in-situ stress, improves the plastic deformation of coal and rock mass, weakens the stress concentration in mining region and finally makes the subsequent seams easier to be extracted [10]. One of the multiple coal seams is suggested to be extracted firstly so long as it is relatively nonhazardous or low-hazardous. For the rock burst mines, the firstly extracted coal seam is usually called liberated seam. The others are called protected seam. As its advantages, application of multi-seam coal mining enlarges gradually in deep coal mines if only the geological conditions are allowable. It is given the higher priority to control rock burst hazard in application [11].

The reason the multi-seam mining method plays an important role in rock burst prevention is that it generates a stress relaxation zone in surrounding rock mass and the protected seam exploitation in stress relaxation zone can be extracted safely. However, in actual production, it is found that the protected seam exploitation in stress relaxation zone is not always as easy and smooth as it is expected. Especially, in deep coal mining, some unnatural accidents occur occasionally during the exploitation in protected seam. These accidents delay mining schedule and injure the workers in the worse cases. Hence, it should be paid attention during mining. We deduced that the fact protected seam extraction in stress relaxation zone gets unsafe is closely related with a certain degree of lateral stress concentration in the interlayer rock stratum between liberated seam and protected seam. Stress relaxation zone in the multi-seam mining structure is determined by the vertical component of principle stress. But the changes of lateral stress in the multi-seam mining structure, especially, in its stress relaxation zone are indeterminate, and also are paid less attention in current researches. Thus, in this study, aiming at the lateral stress changes in interlayer rock stratum, the authors carried out a field investigation in a deep underground coal mine, explored the lateral stress redistribution and interlayer rock failure behavior, discussed the mechanism and impact of lateral stress concentration, which provides the necessary theoretical and practical experience for deep multi-seam coal mining in rock burst prevention.

THREE "ZONES" OF THE MULTI-SEAM COAL MINING STRUCTURE

When multi-seam mining method is used in the prevention of rock burst hazard, the mining region influenced is always divided artificially into several disparate zones based upon the reconstructed stress in the surrounding rock mass of liberated seam extracted [9]. One commonly recognized division is consisted of stress relaxation zone, stress enhancement zone and stress stabilization zone, or other similar definition [12]. The partition is as shown in Figure 1.

Vertical stress in the region that is under and above the extracted coal seam is decreased greatly as the stress transfer [13]. Because of so, this region is called the main stress relaxation zone. Protected seam that locates in this zone is thought to be safely extracted. It is incontrovertible that liberated seam exploitation dramatically lowers the stress concentration degree, weakens the elastic strain energy, and creates positive conditions for the subsequent protected seam mining. Accordingly, rock burst hazard is relieved. With the steeply exploitation in liberated coal seam, dynamic development process of stress redistribution and coal deformation in protected seam wins much attention in current researches, which mainly concentrates on the stress evolution surrounding working face, determination of pressure releasing angle and safe mining area, and optimized layout of roadways [14-16].

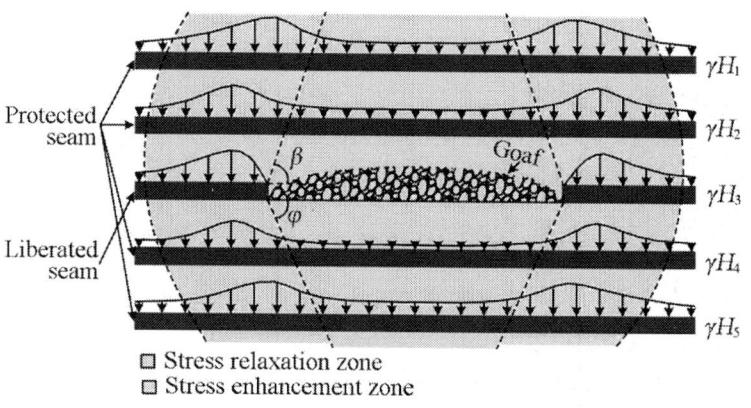

Figure 1: Stress division of the mining region after the liberated seam exploitation. Coal seam in the middle is the liberated seam. Coal seams under and above liberated seam are the protected seams. β and φ are the upper and lower pressure releasing angle, respectively.

FIELD EXPERIMENT OF LATERAL STRESS REDISTRIBUTION IN INTERLAYER ROCK STRATUM

Investigation Site and Scheme

In order to investigate the lateral stress change and redistribution in interlayer rock stratum, a multi-seam coal mining structure in the JNT coal mine, China, was selected as the experiment site. Two coal seams, liberated upper seam (US) and protected lower seam (LS), are worthy of exploitation in this field. Mining depth of upper seam is about −620 m in average. Occurrence thickness of both seams is 2.85 m and 5.0 m. The vertical interlayer spacing is about 42.7 m. Lithology of the interlayer rock mass is the grizzly compact sandstone with a little fracture developed. Its coefficient of hardness is 6 - 8. Face US-04, neighboring the face US-02 to the east and the face US-06 to the west, is located in the central field. It is the first working face extracted. Its mining length and width are 2770 m and 180 m, respectively.

Some connection roadways were tunneled for ventilation requirements during the roadway excavation in protected seam. A stress monitoring hole was drilled in one connection roadway that was ahead of the US-04 working face in horizontal distance. Five stress detectors were installed in the central position then. Lateral stress was monitored during the mining process of Face US-04 until the investigation site was 250 m behind. Meanwhile, four fracture observing holes were drilled with 20 m interval spacing to cooperate with the stress monitoring. Initial fractures of their inner wall were recorded by borehole imaging instrument, and fracture developing situation was observed until the investigation was terminated. Investigation scheme for this field experiment was shown in Figure 2.

Results of Lateral Stress Changes and Interlayer Rock Failure

Observation results of lateral stress redistribution and interlayer rock failure behavior in these boreholes are shown in Figures 3 and 4.

It indicates that lateral stress redistribution in the stress relaxation zone of interlayer rock mass is greatly influenced by the mining activities in liberated seam. When stress detectors are close to but still ahead of the working face (−25 - −10 m), lateral stress in upper rock stratum increases obviously to 32 Mpa but the stress in the lower stratum decreases slightly to 19 Mpa. Lateral stress gets less as rock stratum is further away from working face. When the stress detectors are behind of but still close to the working face (0 - 20 m), lateral stress redistribution reverses dramatically in a short period. Stress in the upper stratum decreases sharply to 16 Mpa, while stress in the lower stratum rises gradually to 26 Mpa. It implies that the lateral stress is concentrating in the low region. While as the stress detectors are behind of and away from the working face (45 - 80 m), lateral stress becomes stable gradually at a high level of 26 Mpa in average. Field investigation results indicate that lateral stress increases in the stress relaxation zone after liberated seam mining, and gradually results in stress concentration in the lower region of interlayer rock stratum. The lower the rock stratum, the greater the degree of stress concentration gets.

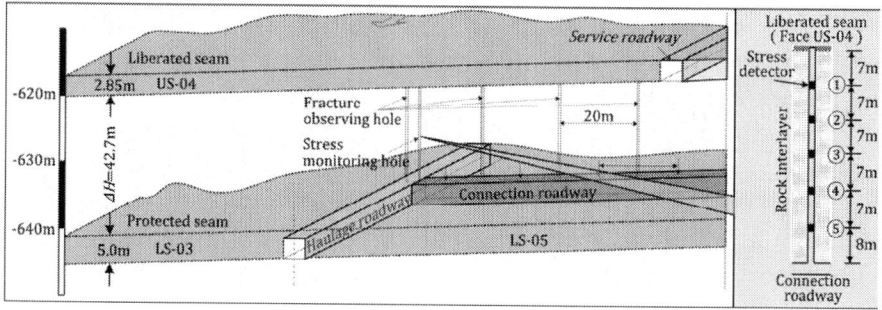

Figure 2: Lateral stress investigation in a multi-seam coal mining structure.

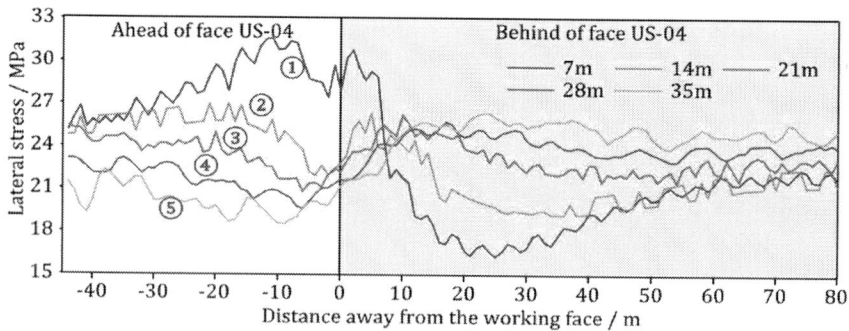

Figure 3: Lateral stress changes in different interlayer rock stratum with the liberated seam exploitation.

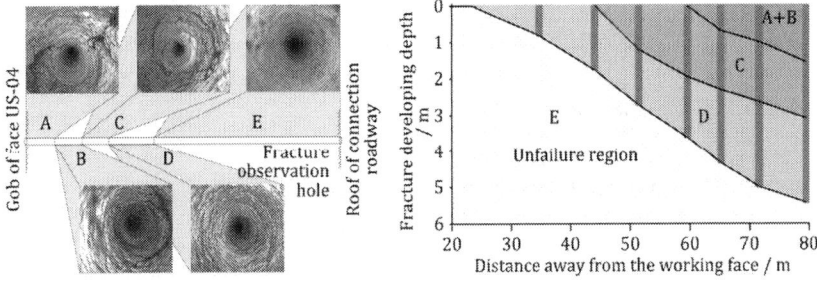

Figure 4: Plastic failure of interlayer rock mass after the liberated seam exploitation.

Beside, plastic failure of interlayer rock mass also presents the distinct manifestation. As the stress reconstruction after upper seam mining, inner wall of boreholes suffers serious destruction. In vertical direction, destruction situation of these four boreholes is similar. Fractures are completely developed in their upper part. The closer to the goaf bottom, the more fractures the boreholes have. In the horizontal direction, the destructtion of borehole nearby the centerline is more serious than others, whereas the destruction in the borehole nearby the haulage roadway is the slightest. And in the mining direction, fracture developing depth increases as the investigation site gets far away from the working face.

NUMERICAL INVESTIGATION ON THE CAUSE OF LATERAL STRESS CONCENTRATION

Construction of Analysis Model

To explore the cause that results in the lateral stress concentration in stress relaxation zone of interlayer rock stratum, numerical analysis method is introduced into this section. Multi-seam coal mining model shown in Figure 5 is built by the finite element analysis software FLAC^{3D} [17]. As it can track the stress changes and failure behavior of analysis object based on the finite difference algorithm and large deformation, this software is particularly appropriate for the researches on rock mass, and has become one of the most important tools in mining engineering. The mechanical properties of coal and rock specimen applied are listed in Table 1.

Stress Redistribution in Interlayer Rock Stratum

Simulated results are shown in Figure 6. It indicates that, as the mining disturbance, in-situ stress reconstructs in a wide region and finally reaches to a new balanced state [18]. In the stress relaxation zone of interlayer rock stratum, vertical stress changes sharply with a great decrement. The ground pressure is released dramatically. However, lateral stress changes following the specific rules. It decreases and is less than the initial value in upper part of interlayer rock, whereas it is obviously greater than the initial value in the lower half region. This case implies that lateral stress increases with the occurrence depth. It leads to a certain extent of stress concentration in the lower rock stratum.

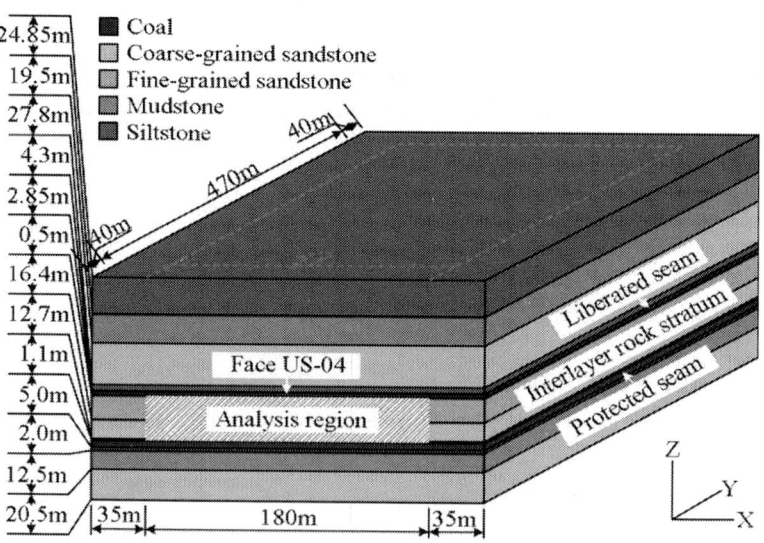

Figure 5: Numerical analysis model of the multi-seam coal mining structure.

Table 1: Basic mechanical properties of coal and rock specimen used in the simulation

	Density kg.tn^{-3}	Elastic modulus GPa	Poisson ratio	Shear modulus GPa	Cohesion Mpa
Coal	13.8	6.1	0.30	2.3	1.0
Mudstone	21.4	11.5	0.25	4.8	4.2
Coarse-grained sandstone	23.5	19.8	0.23	8.5	5.2
Fine-grained sandstone	24.8	20.2	0.23	9.4	5.1
Siltstone	25.7	22.4	0.22	10.2	5.0

Failure State of Interlayer Rock Mass

Simultaneously, the failure state of interlayer rock mass is shown in Figure 7. As the impact of upper seam mining, plastic failure of interlayer rock mass mainly occurs in two regions: the upper one close to the goaf of liberated seam and the lower one in the immediate roof of protected seam. The former is more serious than the latter. Failure behavior breaks the integrity of interlayer rock mass and changes with the redistributed stress [19, 20]. It indicates that numerical results are well coincident with the field experiment results.

Interaction of Lateral Stress Changes and Interlayer Rock Failure

Above discussion indicates that lateral stress concentration and interlayer rock failure are interactional. Failure extent of interlayer rock mass depends on the reconstructed lateral stress. Rock failure phenomenon in the bottom of rock stratum does not mean that the lateral stress transfers to other regions. On the contrary, it happens just as the greater action of lateral stress. Under their interaction, lateral stress itself gradually enhances with the increase of occurrence depth, and finally results in the lateral stress concentration in the lower half region this interaction continues until an equilibrium state is achieved. In a word, lateral stress reconstruction provides mechanical requirements for the failure of interlayer rock mass, and this plastic rock failure behavior further promotes the lateral stress concentration in stress relaxation zone. Lateral stress concentration increases the elastic strain energy in unbroken interlayer rock mass. Once protected seam is extracted, the interlayer rock mass will cave spontaneously as its roof. In this case, large amount of elastic energy release instantly along with the caving process [21, 22]. It undoubtedly induces the hidden accidents and threatens the safe production.

IMPACT OF LATERAL STRESS CONCENTRATION ON PROTECTED SEAM EXPLOITATION

As the lateral stress concentrates on the lower part of interlayer rock, the discussion on its scope and extent then becomes necessary. Hence, another seven models with distinct interlayer spacing, 10 m to 80 m, interval 10 m, were created to explore the impact of lateral stress concentration on protected seam mining, which is assessed by:

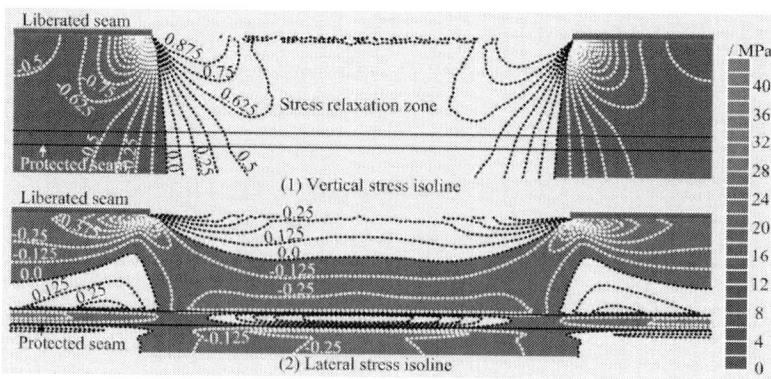

Figure 6: Vertical and lateral stress redistribution in the interlayer rock stratum after liberated seam mining. The value of stress contours is

pressure releasing coefficient, which is calculated by the formula: r = (δ − δ₎)/δ . δ and δ₎ are the initial stress and reconstructed stress, respectively. The less the value, the higher the current stress is.

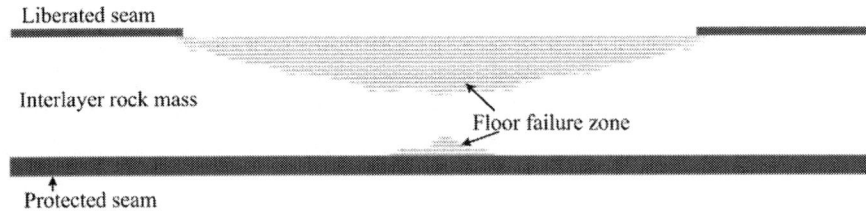

Figure 7: Plastic failure state of interlayer rock mass.

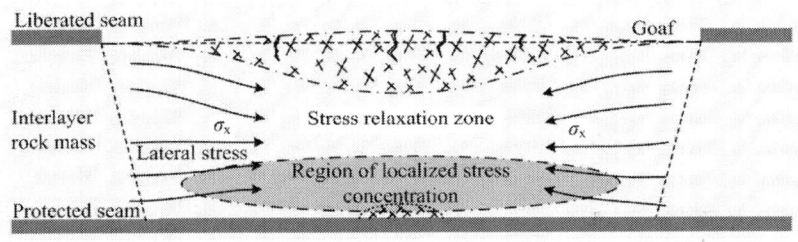

Figure 8: Lateral stress concentration in interlayer rock mass.

$$f_c = \frac{S_l{}'- S_l}{S_l} \times 100\%$$

(1)

$$r_c = \frac{C'_{(f_c>0)}}{C} \times 100\%$$

(2)

where S_l is the initial lateral stress, S_l' is the reconstructed lateral stress; C is the coverage of stress relaxation zone, and C' is the coverage of lateral stress concentration zone; f_c and r_c are the extent and scope of lateral stress concentration, respectively. Lateral stress redistribution results and its concentration scope and degree in distinct interlayer spacing are shown as Figures 9 and 10.

Above figures show that lateral stress is almost completely

concentrated on the whole interlayer rock stratum as interlayer spacing is less than 10 m. When the spacing gets large, the location of stress concentration gradually moves downward and its scope increases. The extent of stress concentration rises little by little and reaches to the maximum as interlayer spacing is 50 m, then decreases slowly. The lateral stress concentration behavior is still obvious even as interlayer spacing is greater than 80 m. Based upon the rules of lateral stress concentration, whether protected seam is easy to be extracted can be divided into three types: safe mining in protected seam as interlayer spacing is less than 35 m, risky mining in protected seam as interlayer spacing is between 35 m and 65 m, and the general mining in protected seam as interlayer spacing is greater than 65 m.

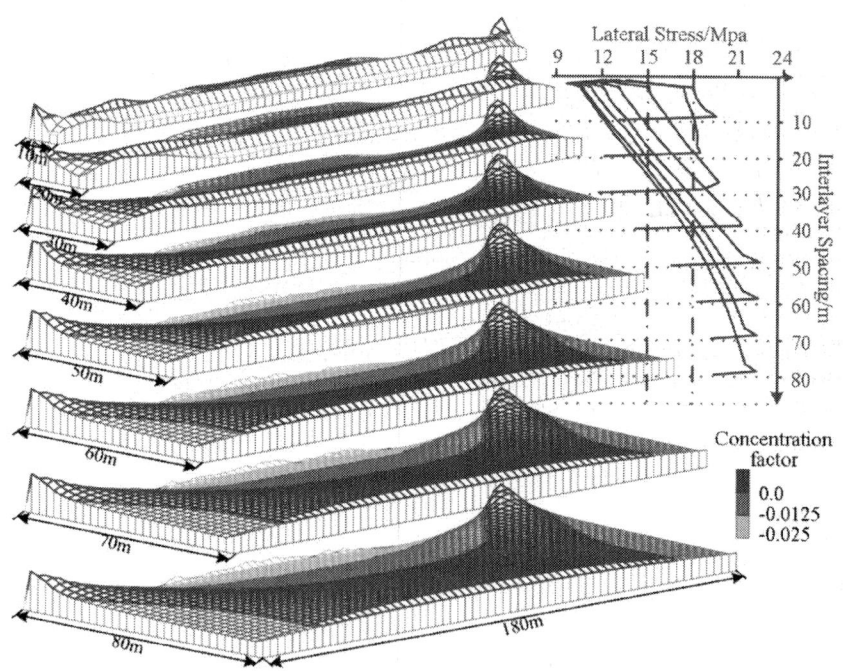

Figure 9: Extent and scope of lateral stress concentration in stress relaxation zone with different spacing.

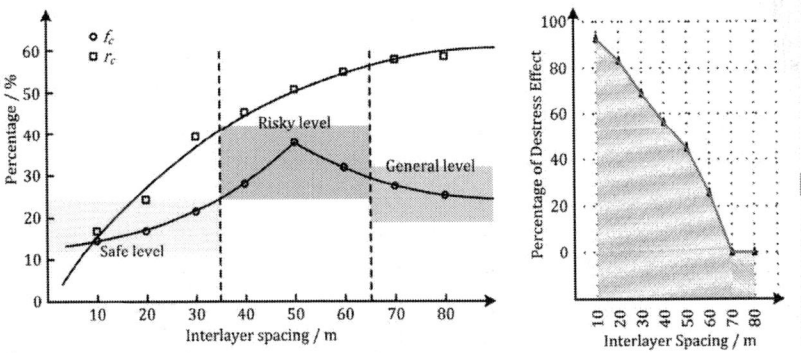

Figure 10: Assessment of lateral stress concentration in distinct interlayer spacing and the corresponding pressure releasing effect.

It indicates that lateral stress concentration in the stress relaxation zone of interlayer rock mass is greatly affected by the interlayer spacing of coal seams. On the one hand, principal stress transfers and reconstructs surrounding the liberated working face; on the other hand, lateral stress itself grows with the increasing interlayer spacing naturally. Besides, the liberated seam mining has limited pressuring releasing efficacy on the exploitation of protected seam, which is closely influenced by the lateral stress concentration in deep multi-seam coal mining. Although protected seam in the stress relaxation zone can be relative securely extracted, lateral stress concentration in the interlayer rock mass certainly brings adverse impact to its extraction. The greater the extent and scope of lateral stress concentration is, the higher the potential threaten to protected seam gets.

CONCLUSIONS

Based on the above discussion about the lateral stress concentration in stress relaxation zone of interlayer rock stratum, the following points can be concluded:

- After liberated seam mining, protected coal seam in its stress relaxation zone is not always safe enough to be extracted as it is generally expected. Decreased vertical stress provides it favorable mechanical environment. However, especially in deep conditions, lateral stress increases instead and can result in a certain extent of stress concentration in the lower half part of interlayer rock stratum. Lateral stress concentration increases elastic strain energy in the unbroken rock mass, and then induces some potential accidents during protected seam mining. This situation is less optimistic for rock burst prevention.

- The cause that induces the lateral stress concentration in stress relaxation zone is closely related with the interlayer rock failure behavior. Lateral stress concentration and interlayer rock failure are interactional. Lateral stress is also the primary inducing factor of unstable interlayer rock failure. In general, the increased lateral stress provides mechanical requirements for the failure process of interlayer rock mass. Meanwhile, interlayer rock failure further promotes the lateral stress concentration in its stress relaxation zone.

- Lateral stress concentration occurs in all of the deep multi-seam mining structure. It is affected by the interlayer spacing between coal seams. The stress concentration extent reaches to the maximum as spacing is 50 m. In this case, coal exploitation in protected seam is the most hazardous. The degree of mining safety in protected seam is divided into three types: safe mining as interlayer spacing is less than 35 m, risky mining as interlayer spacing is 35 - 65 m, and the general mining as interlayer spacing is greater than 65 m. Besides, the pressure releasing effect decreases as the increase of interlayer spacing.

ACKNOWLEDGEMENTS

Financial support for this work provided by the Laboratory of Rock Engineering and Mining Machinery, and the G-COE Program in Novel Carbon Resource Sciences, Kyushu University, is gratefully acknowledged.

REFERENCES

1. B. B. Tati, "Multi-Seam Coal Mining," Journal of The South African Institute of Mining and Metallurgy, Vol. 111, No. 4, 2011, pp. 231-242.

2.   S. Khare, Y. V. Rao, C. S. Murthy and H. Vardhan, "Multiple Seam Mining: A Critical Review," Journal of Mines, Metals and Fuels, Vol. 54, No. 12, 2006, pp. 327- 329.

3.   Z. K. Lin, J. P. Du, J. L. Xu and G. M. Feng, "The Coal Mining Method," China University of Mining and Technology Press, Xuzhou, 2009.

4.   T. V. Lobanova, "Geomechanical State of the Rock Mass at the Tashtagol Mine in the Course of Nucleation and Manifestation of Rock Bursts," Journal of Mining Science, Vol. 44, No. 2, 2008, pp. 146-154. http://dx.doi.org/10.1007/s10913-008-0028-8

5.   L. M. Dou and X. Q. He, "Theory and Technology of Rock Burst Prevention," China University of Mining and Technology Press, Xuzhou, 2001.

6.   L. M. Dou and X. Q. He, "Mining Geophysics," China Science and Culture Press, Beijing, 2002.

7.   R. X. Shen, E. Y. Wang, Z. T. Liu and Z. H. Li, "Rockburst Prevention Mechanism and Technique of Close-Distance Lower Protective Seam Mining," Journal of China Coal Society, Vol. 36, No. S1, 2011, pp. 63-67. (In Chinese)

8.   T. H. Yang, T. Xu, H. Y. Liu, C. A. Tang, B. M. Shi and Q. X. Yu, "Stress-Damage-Flow Coupling Model and Its Application to Pressure Relief Coal Bed Methane in Deep Coal Seam," International Journal of Coal Geology, Vol. 86, No. 4, 2011, pp. 357-366.http://dx.doi.org/10.1016/j.coal.2011.04.002

9.   X. Q. Wu, L. M. Dou, C. G. Lv, A. Y. Cao and M. W. Zhang, "Research on Pressure-Relief Effort of Mining Upper-Protective Seam on Protected Seam," Procedia Engineering, Vol. 26, 2011, pp. 1089-1096. http://dx.doi.org/10.1016/j.proeng.2011.11.2278

10.   D. A. Beck and B. H. Brady, "Evaluation and Application of Controlling Parameters for Seismic Events in HardRock Mines," International Journal of Rock Mechanics and Mining Sciences, Vol. 39, No. 5, 2002, pp. 633-642. http://dx.doi.org/10.1016/S1365-1609(02)00061-8

11.   Ministry of Coal Industry, "National Provisional Rules of Safe Mining in Rock Burst Seam," 1987.

12. W. Yang, B. Q. Lin, Y. A. Qu, Z. W. Li, C. Zhai, L. L. Jia and W. Q. Zhao, "Stress Evolution with Time and Space during Mining of a Coal Seam," International Journal of Rock Mechanics and Mining Sciences, Vol. 48, No. 7, 2011, pp. 1145-1152.http://dx.doi.org/10.1016/j.ijrmms.2011.07.006

13. B. H. G. Brady and E. T. Brown, "Rock Mechanics for Underground Mining," Kluwer Academic Publishers, 2004.

14. W. Yang, B. Q. Lin, Y. A. Qu, S. Zhao, C. Zhai, L. L. Jia and W. Q. Zhao, "Mechanism of Strata Deformation under Protective Seam and Its Application for Relieved Methane Control," International Journal of Coal Geology, Vol. 85, No. 3-4, 2010, pp. 300-306. http://dx.doi.org/10.1016/j.coal.2010.12.008

15. J. C. Wang and H. B. Zhao, "Numerical Simulation on Deformation Rule of Protected Coal Seam under Upper Protective Seam Method," Disaster Advances, Vol. 3, No. 4, 2010, pp. 383-387.

16. Y. K. Liu, F. B. Zhou, L. Liu, C. Liu and S. Y. Hu, "An Experimental and Numerical Investigation on the Deformation of Overlying Coal Seams above Double-Seam Extraction for Controlling Coal Mine Methane Emissions," International Journal of Coal Geology, Vol. 87, No. 2, 2011, pp. 139-149. http://dx.doi.org/10.1016/j.coal.2011.06.003

17. N. Z. Xu and L. Han, "Pressure-relief Effect of Coal Rock Body of Long Distance Lower Protective Seam Mined Based on FLAC3D," Journal of Coal Science and Engineering, Vol. 16, No. 4, 2010, pp. 341-346. http://dx.doi.org/10.1007/s12404-010-0402-2

18. H. Guo, L. Yuan, B. T. Shen, Q. D. Qu and J. H. Xue, "Mining-Induced Strata Stress Changes, Fractures and Gas Flow Dynamics in Multi-Seam Longwall Mining," International Journal of Rock Mechanics and Mining Sciences, Vol. 54, 2012, pp. 129-139. http://dx.doi.org/10.1016/j.ijrmms.2012.05.023

19. A. A. Nasedkina, A. V. Nasedkin and G. Iovane, "A Model for Hydrodynamic Influence on a Multi-Layer Deformable Coal Seam," Computational Mechanics, Vol. 41, No. 3, 2008, pp. 379-389. http://dx.doi.org/10.1007/s00466-007-0194-6

20. S. Prusek and S. Bock, "Assessment of Rock Mass Stresses and Deformations around Mine Workings Based on Three-Dimensional Numerical Modeling," Archives of Mining Sciences, Vol. 53, No. 3, 2008, pp. 349-360.

21. Z. Q. Yin, X. B. Li, J. F. Jin, X. Q. He and K. Du, "Failure Characteristics of High Stress Rock Induced by Impact Disturbance under Confining Pressure Unloading," Transactions of Nonferrous Metals Society of China, Vol. 22, No. 1, 2012, pp. 175-184.http://dx.doi.org/10.1016/S1003-6326(11)61158-8

22. R. Weinberger, Y. Eyal and N. Mortimer, "Formation of Systematic Joints in Metamorphic Rocks Due to Release of Residual Elastic Strain Energy, Otago Schist, New Zealand," Journal of Structural Geology, Vol. 32, No. 3, 2010, pp. 288-305.http://dx.doi.org/10.1016/j.jsg.2009.12.003

Gold Mining and Mercury Bioaccumulation in a Floodplain Lake and Main Channel of the Tambopata River, Perú

Katherine A. Roach[1], Nicolas F. Jacobsen[2], Christine V. Fiorello[3], Amanda Stronza[2], and Kirk O. Winemiller[1]

[1]Department of Wildlife and Fisheries Sciences, Texas A&M University, College Station, USA

[2]Department of Recreation, Park and Tourism Sciences, Texas A&M University, College Station, USA

[3]Wildlife Health Center, School of Veterinary Medicine, University of California, Davis, USA

ABSTRACT

Contamination of water bodies by inorganic mercury (Hg[II]) used in placer mining of gold deposits in the Madre de Dios Department, Perú, contributes to the bioaccumulation of methylmercury (MeHg) in fish tissue. We measured MeHg and total Hg (THg) concentrations (mg/kg wet weight [ww] tissue) of thirteen fish species from the Tambopata River, Perú, and the connected oxbow lake Tres Chimbadas. We also used stable isotope analysis (^{15}N and ^{13}C) to estimate trophic positions of fishes. Average MeHg concentrations of fish species ranged from 0.042 mg/kg ww (Satanoperca jurupari) to 0.463 mg/kg ww (Hoplias malabaricus) in the main channel and from 0.090 mg/kg ww (Parauchenipterus sp.) to 1.282 mg/kg ww (Pimelodina flavipinnis) in the lake. Spearman rank correlation indicated that trophic position had no influence on MeHg concentrations of species in the main channel, but in the lake, trophic positions of species were positively associated with MeHg. Migrations of the pimelodid catfish surveyed from the main channel are well documented. Because little gold mining occurs at our study site, fishes from the main channel may be bioaccumulating MeHg from areas where mining is widespread. Fish species that reside in the lake are relatively sedentary and migration is limited by the brief period of floodplain inundation and the long, narrow corridor that connects the lake to the main channel; lake sediments are therefore the likely source for MeHg bioaccumulation. Five out of the eight fish species surveyed from the main channel and two out of the five species from the lake had MeHg levels higher than United States Environmental Protection Agency fish tissue criterion for human consumption.

INTRODUCTION

Mercury (Hg) is a pollutant that can cause developmental and behavioral abnormalities including impaired reproduction and decreased survival in vertebrates [1]. Methylmercury (MeHg) is the form of most concern because, unlike inorganic Hg (Hg[II]) which is readily excreted from the body, MeHg bioaccumulates and can reach harmful levels in fish tissues [2]. The US Environmental Protection Agency estimates (with uncertainty spanning an order of magnitude) that Hg exposure of the human population, including sensitive subgroups, of 0.0001 mg/kg/day is likely to be without appreciable risk of deleterious effects during a lifetime [3]. Based on this estimate, the USEPA recommends that 0.3 mg MeHg/kg wet weight (ww) tissue should not be exceeded to protect the health of consumers of noncommercial freshwater and estuarine fish [3].

In the Madre de Dios department, Perú, placer mining of gold deposits is believed to have contributed to increased Hg(II) concentrations in the water and sediments of some rivers. This is of concern for local people because they consume large amounts of fish from local rivers. Traditionally, fish were the most significant source of animal protein for local people, and while diets have diversified due to an increase in ranching as well as connection to national and global markets, locally caught fish remain an important part of the diet. The majority of gold mining operations in Madre de Dios are relatively small scale (e.g., family-operated placer mines), but in recent years both the number and scale of these operations have increased [4]. With the recent rise in worldwide gold prices (currently >$1500/ounce), there is economic pressure to expand this industry [5,6]. Hg(II) is used to amalgamate gold from river sediment, and during this process up to 20% of the Hg used is lost into the river. Another 20% can then be lost to the atmosphere in the subsequent process of heating the metal to separate the Hg from the gold [7]. Because estimates of Hg pollution from gold mining are imprecise, little is known about sources and sinks for Hg in Madre de Dios watersheds. Oxbow lakes in this region have been hypothesized to retain Hg because they have high sediment organic matter content [8]. Impacts from placer gold mining operations in this region also have been suggested to be limited to river reaches immediately downstream of mining sites [9], which implies that fish with high concentrations of MeHg do not move very far from the source of contamination. Given the great extent of mining activity and the long-distance movements of many of the fish species consumed by humans in this region [10], Hg pollution from gold mining poses a potentially serious risk to human health.

Humans are not the only species affected by Hg pollution and bioaccumulation; the entire aquatic ecosystem can be negatively impacted. At MeHg dietary concentrations of 0.5 mg/kg ww, many fish species experience impaired reproduction, and significant mortality occurs at concentrations >1.9 mg/kg ww [11]. A species of particular concern is the giant otter, Pteronura brasiliensis, listed as endangered by the International Union for Conservation of Nature [12]. A flagship species for both conservation and ecotourism in the Madre de Dios region, giant otters are an apex predator in the aquatic food web residing in rivers and, more frequently, lakes. Their diet consists almost exclusively of fish, and many of their preferred prey species occupy relatively high trophic positions [4,13]. In addition, giant otters have high metabolic rates; to support this energy consumption, they consume up to 4 kg of fish per day [4]. While very little is known about the impacts of MeHg accumulation on giant otters of South America, studies indicate that mammalian wildlife and humans respond to MeHg in a comparable manner [14]. Furthermore, other members of the family Mustelidae including otter species in North America and Europe (Lutra canadensis, Lutra lutra) and American mink (Mustela vison) are highly sensitive to Hg, with neurochemical changes in the brain occurring at MeHg dietary concentrations of 0.5 mg/kg [15-17].

The aquatic Hg cycle is complex, with many interrelated factors determining fish MeHg concentrations. The fundamental drivers of this process include Hg deposition, the transformation of inorganic Hg(II) to MeHg by sulfate-reducing bacteria under anoxic environmental conditions [18], and rates of MeHg uptake and trophic transfer [19,20]. Comparisons of species' trophic positions and MeHg concentrations can allow for inferences to be made about the factors that influence MeHg bioaccumulation. Nitrogen stable isotope ratios (^{15}N) can be used to estimate consumer trophic positions because tissue becomes more enriched in ^{15}N with each subsequent trophic transfer ["fractionation", 21,22]. However, the ^{15}N signature of primary producers can vary among habitats, thus a baseline ^{15}N must first be taken into account. Because ^{13}C does not fractionate up the food chain, it is useful for estimating the basal production sources (i.e., algae, terrestrial plants) from which a consumer derives its baseline ^{15}N.

Our major objectives of the present study were to measure total Hg (THg) and MeHg concentrations and trophic positions of thirteen fish species, including six species that are important food resources for humans and giant otters, in the Tambopata River and the oxbow lake Tres Chimbadas, Perú to determine if Hg levels were above USEPA fish tissue criterion. We then examined how the relationship between trophic position and MeHg of fish species varied between the main channel and oxbow lake. We also measured Hg(II) concentrations and organic matter content of sediment in both habitats. Finally, we compared our results with those of other studies of Hg concentrations in fish from aquatic ecosystems in the Madre de Dios Department, Perú.

STUDY AREA

The Tambopata River (**Figure 1**) is a meandering, socalled whitewater river (i.e., carries high suspended sediment loads) that originates in the Andean piedmont of Perú, meets the Madre de Dios River in Perú, and meets the Amazon River in Brazil. The hydrologic regime shows a distinct seasonal flow pattern; stage height is generally highest from February to March and lowest from August to September, however rapid fluctuations in stage height of 2 - 3 m per day can occur at any time of the year [23]. Suspended sediment load is high due to erosion of the Andes, and the low-gradient, deep alluvial sediments, and variable discharge produce lateral channel migrations [24]. A small stream flows from the oxbow lake Tres Chimbadas to the main channel. In our study region, the floodplain mostly consists of intact moist tropical forest, but smallscale farming and cattle ranching also occur. Tres Chimbadas supports giant otters and was designated a conservation reservation by the local Ese'eja native community of Infierno, thus neither mining nor fishing for consumptive purposes are permitted [25].

MATERIALS AND METHODS

Sample Collections

From May to August 2009, fishes were collected from the main channel and lake with seines, cast nets, and hook and line. Following euthanasia, an approximately 5 - 10 g sample of muscle tissue from each individual was removed from the dorso-lateral region using a scalpel. Subsamples of tissue for analysis of Hg were placed in acidwashed bottles filled with ethanol, and sub-samples for stable isotope analysis were placed in a plastic bag with salt, a preservation method that has minimal influence on stable isotope signatures [26]. A minimum of 100 g of bed sediment was collected using a shovel and hand trough. A surficial sample was taken first, sealed in a plastic bag, and labeled. A shovel was then used to dig at least 30 cm deep or until the color and texture of the sediment changed. At this point, a deep sediment sample was collected in the same manner. For each location, this process was repeated at two additional sites, each separated by at least 20 m.

Samples for Hg analysis were preserved by freezing. Sediment surface samples for analysis of organic content were collected by inverting a Petri dish (5-cm diameter and 1.3-cm height), pressing the edge into the sediment, and capping it with a spatula. These samples were preserved in 10% formalin. Samples for Hg analysis were transported to the University of Georgia, and samples for stable isotope analysis and sediment organic matter content were transported to Texas A&M University for processing.

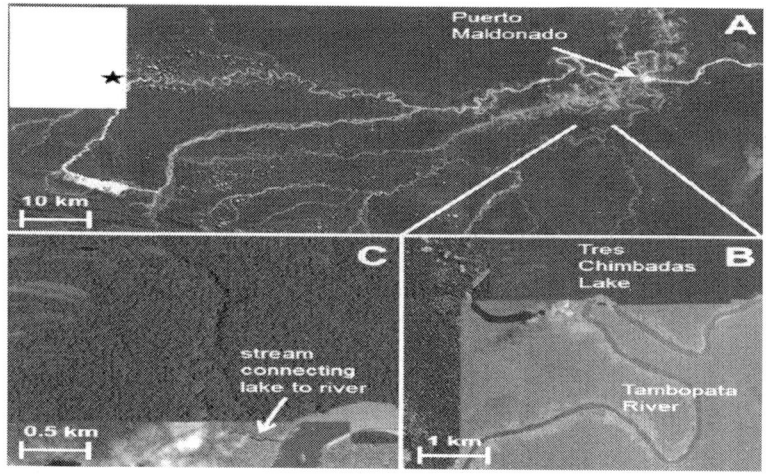

Figure 1: (A) Map of the study site ca. 25 km southwest of Puerto Maldonado, Perú. Changes in geomorphology of the Inambari River as a result of placer mining can be observed on the bottom left of the photograph; (B) Tres Chimbadas Lake and the Tambopata River main channel where samples were collected for mercury and stable isotope analysis. Latitude (°S)/longitude (°W) of habitats: Tres Chimbadas Lake, 12.47.2246/69.20.4462; Tambopata River main channel, 12.48.0406/69.17.3063; (C) The small stream that permanently connects Tres Chimbadas to the main channel.

Analytical Methods

Each set of triplicate sediment samples was pooled prior to Hg analysis. Samples of fish muscle and sediment for Hg analysis were sent to a commercial analytical laboratory (Quicksilver Scientific, Lafayette, CO,

USA) for analysis using high pressure liquid chromatography speciation analysis (QS-LC/CVAF-001) to determine Hg(II) and MeHg [27]. Several quality control measures were taken during the analysis, including 2 blanks, 2 reference materials, 2 matrix spike duplicates, 2 laboratory control samples, and 2 sample duplicates. Method detection limits for fish tissue were 175.5 pg/g MeHg and 445.0 pg/g Hg(II). Recovery of control samples was 102.0% for MeHg and 100.3% for THg. The average relative percent difference in THg of duplicate samples was 5.55%. We assumed 80% moisture in fish tissue samples [28] and thus a conversion factor of 5 was used to convert dry to wet weight in order to compare samples with USEPA guidelines.

At the laboratory at Texas A&M University, samples for stable isotope analysis were rinsed and soaked for 4 h in deionized water to remove salt. Samples were dried at 65°C for 48 h, ground to a fine powder using a mortar and pestle, and weighed into Ultra-Pure tin capsules (Costech Analytical, Valencia, CA, USA). Samples were subsequently sent to the W.M. Keck Paleoenvironmental and Environmental Stable Isotope Laboratory, University of Kansas, Lawrence, KS, USA, for analysis of carbon and nitrogen isotope ratios on a ThermoFinnigan MAT 253 continuous-flow mass spectrometer using standards of Pee Dee Belemnite limestone for carbon isotopes and atomspheric nitrogen for nitrogen isotopes. We measured sediment organic matter content using the % ash-free dry mass method [29].

Data Analysis

Trophic position estimates were calculated based on a standard two-source mixing model using the equation

$$TP = \lambda + \left(\delta^{15}N_{sc} - \left[\left(\delta^{15}N_{base1} \times \alpha + \delta^{15}N_{base2} \times (1-\alpha) \right) \right] \right) / 2.54$$

where λ was the trophic level of the food base (2 for primary consumers), α was the proportion of ^{15}N in the consumer derived from base 1, and 2.54‰ was the average trophic fractionation value [21,22]. We estimated α using the equation

$$\left(\delta^{13}C_{sc} - \delta^{13}C_{base2} \right) / \left(\delta^{13}C_{base1} - \delta^{13}C_{base2} \right)$$

Because primary consumers integrate the temporal and spatial variability in stable isotope signatures of primary producers [21], we selected two common primary consumers to estimate the baseline. The fish species Prochilodus nigricans (Characiformes: Prochilodontidae) assimilates autochthonous algae and detritus, and was present in both the lake and main channel. We selected two fish species that assimilate terrestrial plants: Astyanax abramoides (Characiformes: Characidae) from the lake and Brycon amazonicus (Characiformes: Characidae) from the main channel.

Spearman rank correlation was used to examine the relationship between average trophic positions and average MeHg concentrations of fish species in each habitat. Student's t-tests were used to examine between-habitat differences in sediment organic matter content. Statistical analyses were performed in SPSS, and results were considered statistically significant at = 0.05.

RESULTS

Hg Concentrations

THg in individual fishes ranged from 0.035 to 1.351 mg/kg ww tissue, and MeHg ranged from 0.035 to 1.282 mg/kg ww tissue. Fishes from the main channel generally had higher THg (range 0.095 to 1.351 mg/kg ww tissueaverage of all species surveyed = 0.586 mg/kg ww tissue) and MeHg (range 0.090 to 1.282 mg/kg ww tissue, average for all species surveyed = 0.562 mg/kg ww tissue) concentrations than fishes from the lake (THg range 0.043 to 0.466 mg/kg ww tissue and average for all species surveyed = 0.219 mg/kg ww tissue; MeHg range 0.042 to 0.463 mg/kg ww tissue and average for all species surveyed = 0.217 mg/kg ww tissue; Table 1).

MeHg concentrations in some fishes from both the lake and the main channel were above USEPA fish tissue criterion. In the lake, piraña (Serrasalmus sp., 0.304 mg/kg ww tissue) and huasaco (Hoplias malabaricus, 0.463 mg/kg ww tissue) were above USEPA fish tissue criterion, and in the main channel, toa (Platystomatichthys sturio, 0.590 mg/kg ww tissue), zungaro (Zungaro zungaro, 0.642 mg/kg ww tissue), maparate gordo (Ageneiosus brevifilis, 0.733 mg/kg ww tissue), mota fina (Pinirampus pirinampu, 0.974 mg/kg ww tissue), and mota con puntos (Pimelodina flavipinnis, 1.282 mg/kg ww tissue) were above

USEPA criterion.

Table 1: Average (±SD) total length (cm), methylmercury (MeHg, mg/kg wet weight tissue), and total mercury (THg, mg/kg wet weight tissue) concentrations in muscle tissue of fishes from the Tambopata River main channel and Tres Chimbadas oxbow lake, Perú. Common names of fishes and sample sizes are in parentheses next to species taxonomic names

Taxa	Total length (cm)	Habitat	MeHg	THg
Characiformes				
Characidae				
Brachychalcinus sp. (mojarita, 1)	6.3	Lake	0.053	0.054
Serrasalmus sp. (piraña, 2)	15.5 (0.5)	Lake	0.304 (0.138)	0.320 (0.142)
Triportheus angulatus (sapamama, 1)	12.1	Lake	0.223	0.223
Erythrinidae				
Hoplias malabaricus (huasaco, 2)	22.4 (5.9)	Lake	0.463 (0.381)	0.466 (0.383)
Perciformes				
Cichlidae				
Satanoperca jurupari (bujurqui, 2)	15.2 (1.0)	Lake	0.042 (0.011)	0.043 (0.011)
Gymnotiformes				
Sternopygidae				
Sternopygus macrurus (macana, 1)	32.0	Main channel	0.124	0.126
Siluriformes				
Auchenipteridae				
Agenetosus brevifilis (maparate gordo, 1)	50.8	Main channel	0.733	0.713
Parauchenipterus sp. (1)	19.0	Main channel	0.090	0.095
Pimelodidae				
Pimelodina flavipinnis (mota con puntos, 1)	64.8	Main channel	1.282	1.351
Pimelodus blochii (bagre, 1)	20.2	Main channel	0.091	0.092
Pinirampus pirinampu (mota fina, 1)	68.6	Main channel	0.974	1.030
Platystomatichthys sturio (toa, 1)	32.0	Main channel	0.590	0.602
Zungaro zungaro (zungaro, 1)	92.7	Main channel	0.642	0.682

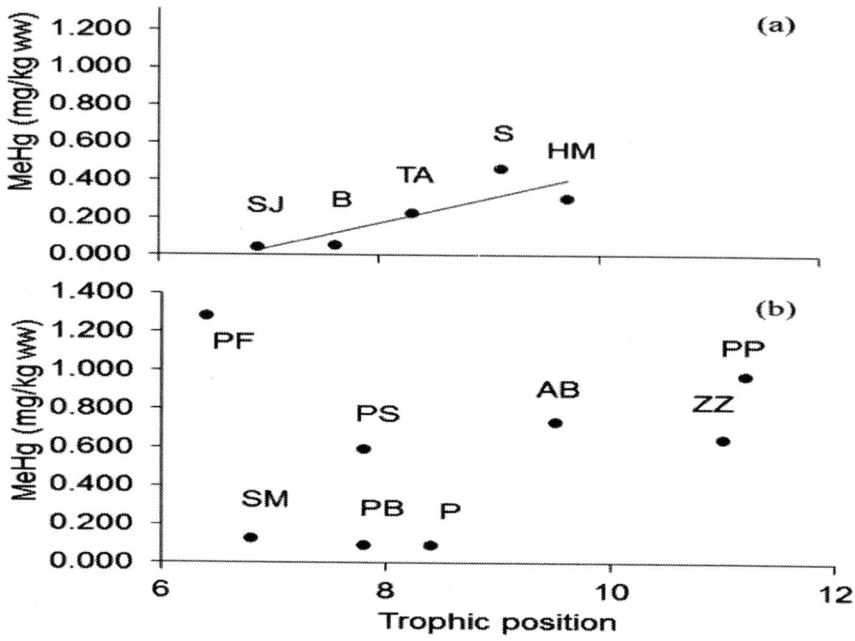

Figure 2: Relationship between average methylmercury (MeHg) muscle tissue concentrations (mg/kg wet weight) and trophic positions of fishes from (a) Tres Chimbadas Lake (SJ = Satanoperca jurupari, B = Brachychalcinus sp., TA = Triportheus angulatus, S = Serrasalmus sp., HM = Hoplias malabaricus) and (b) the Tambopata River main channel (P = Parauchenipterus sp., PB = Pimelodus blochii, SM = Sternopygus macrurus, PS = Platystomatichthys sturio, ZZ = Zungaro zungaro, AB = Ageneiosus brevifilis, PP = Pinirampus pirinampu, and PF = Pimelodina flavipinnis). The regression line (y = ax + c) is shown only for the statistically significant correlation obtained for plot (A).

Trophic Positions

In both habitats, the fishes with the highest MeHg concentrations were not feeding at the highest trophic positions (**Figure 2**). For example, in the main channel mota con puntos had MeHg concentrations twofold higher than zungaro, however zungaro fed at a higher trophic position. Spearman rank correlation indicated that trophic position had no influence on MeHg in the main channel ($r_s = 0.12$, $p = 0.78$). In the lake, species trophic positions were positively associated with MeHg ($r_s = 0.90$, $p < 0.05$; Figure 2).

Sediment Hg (II) and Organic Matter Content

Hg(II) concentration of sediment from Tres Chimbadas lake was higher (22 ng/g ww) compared to sediment from the Tambopata River main channel (13 ng/g ww).

The high organic matter content of the lake sediment (average = 10.83%) indicates that anoxia likely occurs. In contrast, average sediment organic content was only 0.54% in the main channel, probably due to resuspension and downstream transport. Differences in organic matter content were significant between the main channel and the lake (Student's t-test, t = 6.44, df = 25, p < 0.001).

DISCUSSION

Most of the fishes from our study site had Hg concentrations within the range of concentrations found in fishes sampled from the Madre de Dios River near Puerto Maldonado; Rio de la Torre, a tributary of the Tambopata River; and the Puerto Maldonado fish market from 1990-1993 [8], and from the Puerto Maldonado fish market in 2008 [30] (Table 2). One fish from our study site, the pimelodid catfish mota con puntos, had Hg concentrations higher than levels reported previously for fishes from the Madre de Dios department. Three of the five fish species sampled by [8] within 100 km of the city of Puerto Maldonado, where mining of gold deposits is fairly common, were above the USEPA fish tissue criterion. Assuming that 95% of the THg present is MeHg [31], MeHg concentrations documented by [30] ranged from below detection for the characid Piaractus brachypomus to 1.072 mg/kg ww tissue for the Pimelodid catfish Calophysis macropterus. In contrast, fishes sampled by [8] from aquatic habitats inside Manu National Park from 1990-1993 seemed to have slightly lower Hg concentrations; two of the seven fish species sampled had MeHg concentrations above USEPA fish tissue criterion. Again assuming that 95% of THg is MeHg, fishes sampled in 1997 by [32] from an isolated site with no gold mining and very low human population density on the Candemo River, Bahuaja-Sonene National Park, had much lower MeHg concentrations ranging from 0.012 mg/kg ww tissue for the detritivore Prochilodus nigricans to 0.086 mg/kg ww tissue for Pseudoplatystoma sp., a carnivorous pimelodid catfish.

Potential sources of Hg (II) contamination at our study site include naturally occurring Hg from erosion of the Andes Mountains and precipitated Hg particles formed by atmospheric fumes or tailings from gold mining [10]. For example, cinnabar, the most common Hg ore, has been mined in the town of Huancavelica located in the Peruvian Andes for over 3000 years [33]. However, an enormous amount of Hg (II) has been used for gold extraction in the Madre de Dios Department in recent years; Peruvian Hg imports in 2009 were estimated at 175 tons [5], and more than 95% of imports are used for artisanal mining [34]. Additionally, the Madre de Dios Department has the greatest number of unapproved mining permits in Peru (i.e., final approval of a mining permit requires an environmental impact report, [35]) and produces 70% of the country's gold [34]. The high levels of imported Hg combined with the higher Hg concentrations observed in fish collected near mined sites [8,30] compared to those captured near more pristine sites, such as the Candemo River [32], suggest that mining is the most important source of contamination at our study site.

Table 2: Average methylmercury (MeHg, mg/kg wet weight tissue) and total mercury (THg, mg/kg wet weight tissue) concentrations of fishes from the Madre de Dios Department, Perú

System	Year sampled	Family	Genus Species	MeHg	THg	Refere
Fish market, Puerto Maldonado	2008	Characidae	*Hydrolycus pectoralis*	N/A	0.585	[30]
Fish market, Puerto Maldonado	2008	Characidae	*Piaractus brachypomus*	N/A	bd	[30]
Fish market, Puerto Maldonado	2008	Curimatidae	*Potamorhina altamazonica*	N/A	0.021	[30]
Fish market, Puerto Maldonado	2008	Loricariidae	*Pterygoplichthys* sp.	N/A	0.321	[30]
Fish market, Puerto Maldonado	2008	Pimelodidae	*Calophysus macropterus*	N/A	1.128	[30]
Fish market, Puerto Maldonado	2008	Pimelodidae	*Pimelodus* sp.	N/A	0.120	[30]
Fish market, Puerto Maldonado	2008	Pimelodidae	*Pseudoplatystoma fasciatum*	N/A	0.321	[30]
Fish market, Puerto Maldonado	2008	Pimelodidae	*Pseudoplatystoma tigrinum*	N/A	0.183	[30]
Fish market, Puerto Maldonado	2008	Pimelodidae	*Zungaro zungaro*	N/A	0.698	[30]
Fish market, Puerto Maldonado	2008	Prochilodontidae	*Prochilodus nigricans*	N/A	0.036	[30]
Fish market, Puerto Maldonado	2008	Sciaenidae	*Plagioscion squamosissimus*	N/A	0.095	[30]
Candamo River	1997	Characidae	*Brycon melanopterus*	N/A	0.044	[32]
Candamo River	1997	Erythrinidae	*Hoplias malabaricus*	N/A	0.031	[32]
Candamo River	1997	Pimelodidae	*Pimelodus ornatus*	N/A	0.042	[32]
Candamo River	1997	Pimelodidae	*Pseudoplatystoma* sp.	N/A	0.091	[32]
Candamo River	1997	Prochilodontidae	*Prochilodus nigricans*	N/A	0.013	[32]
Cumeriali River	1990-1993	Pimelodidae	*Zungaro zungaro*	0.235	0.281	[8]
Manu River by Tayakome	1990-1993	Characidae	*Piaractus brachypomus*	0.043	0.053	[8]

Manu River by Tayakome	1990-1993	Pimelodidae	*Pseudoplatystoma* sp.	0.068	0.086	[8]
Manu River by Tayakome	1990-1993	Serrasalmidae	*Serrasalmus* sp.	0.069	0.090	[8]
Manu River 50 km downstream of Tayakome	1990-1993	Characidae	*Piaractus brachypomus*	0.057	0.067	[8]
Manu River 50 km downstream of Tayakome	1990-1993	Pimelodidae	*Zungaro zungaro*	0.238	0.280	[8]
Manu River 50 km downstream of Tayakome	1990-1993	Serrasalmidae	*Serrasalmus rhombeus*	0.088	0.140	[8]
Madre de Dios River 30 km upstream of Boca Manu	1990-1993	Curimatidae	*Potamorhina altamazonica*	0.068	0.093	[8]
Madre de Dios River 30 km upstream of Boca Manu	1990-1993	Pimelodidae	*Pseudoplatystoma* sp.	1.071	1.113	[8]
Madre de Dios River 30 km upstream of Boca Manu	1990-1993	Sciaenidae	*Plagioscion auratus*	0.424	0.455	[8]
Manu River 25 km upstream of Boca Manu	1990-1993	Pimelodidae	*Zungaro zungaro*	0.057	0.072	[8]
Manu River 25 km upstream of Boca Manu	1990-1993	Pimelodidae	*Pimelodus* sp.	0.034	0.051	[8]
Madre de Dios River 50 km upstream of Puerto Maldonado	1990-1993	Curimatidae	*Potamorhina altamazonica*	0.038	0.058	[8]
Fish market, Puerto Maldonado	1990-1993	Cynodontidae	*Cynodon* sp.	0.587	0.628	[8]
Fish market, Puerto Maldonado	1990-1993	Pimelodidae	*Zungaro zungaro*	0.770	1.008	[8]
Fish market, Puerto Maldonado	1990-1993	Prochilodontidae	*Prochilodus nigricans*	0.096	0.125	[8]
Rio de la Torre	1990-1993	Pimelodidae	*Pseudoplatystoma* sp.	0.366	0.467	[8]

The gold mining that we have observed in the Tambopata River main channel occurs at a small scale; we saw only one operation during our fieldwork from May to July 2009 near the ecotourism lodge Posada Amazonas. In addition, four or five dredges commonly operate below the confluence of the Malinowski River during the dry season (personal communications, Dr. Donald Brightsmith, 2 May 2011). In contrast, gold mining in neighboring river basins, such as the Colorado and Inambari Rivers, operate at a much larger scale. Although the operations are still considered "artisanal", heavy machinery such as front-loading tractors and dredgers has caused vast changes in river geomorphology that can be observed in satellite images (Figure 1) [23]. The small scale of gold mining that occurs at our study site, high concentrations of sediments in the river that readily bind to Hg (II) [36], and high turnover of sediment associated with scouring floods are potential mechanisms that may explain why sediment Hg concentrations in the main channel are relatively low.

Because of its protected status as a conservation reservation, the oxbow lake Tres Chimbadas is not subjected to gold mining. The lake must either receive Hg particles from the atmosphere or from the main channel via a small connecting stream. The higher Hg(II) concentrations in lake sediments compared to main channel sediments suggest that because suspended sediment concentrations and turnover rate are much lower in the lake, there is a greater potential for Hg to accumulate. Sediment organic matter content also was greater in the lake compared to the main channel, and lake sediments likely are anaerobic as a consequence. Because Hg methylation in sediments requires anaerobic conditions, high sediment organic matter content can promote the production of MeHg by microbes (e.g., [37]). However, organic molecules can bind with Hg, limiting availability of Hg for methylating bacteria [38]. Thus, MeHg tends to not be transferred to the metazoan food web in aquatic habitats with high sediment organic matter content [39-41].

Trophic position is an important consideration in Hg studies because it influences dietary exposure to MeHg. Concentrations of MeHg in fish tissue did not increase with trophic position in the main channel, but trophic position was positively associated with MeHg concentrations in the lake. A strong relationship between MeHg and trophic position would require that consumers assimilate MeHg in food chains derived from the same source of contamination. Long-distance movements by consumers could weaken the correlation between MeHg body burden and trophic position. Seasonal, long-distance migrations of fishes are commonplace in Neotropical rivers [42]. Some of the large predatory species of pimelodid catfishes are believed to move to the Andean foothills to spawn during the beginning of the high-water period in November or December, and then migrate downstream to deeper river channels at the beginning of the low-water period in June [10,43]. In the main channel, some pimelodid species that fed at relatively low trophic positions had high MeHg concentrations (e.g., Parauchenipterus sp. and mota con puntos). These catfishes could have assimilated MeHg from more contaminated rivers in neighboring drainage basins and then migrated into our study area.

In contrast to fishes from the main channel, the fishes captured from the lake seem to be relatively sedentary species. We seined sandbanks in the main channel several times a week from May to July 2009 and August 2010 and captured no fishes from the families Erythrinidae, Cichlidae, or Serrasalmidae, but these families were well-represented in our lake samples. Other studies also have reported species from these families to be common in oxbow lakes [10,44]. Furthermore, because of the relatively high elevational gradient at our study site, river water level rarely exceeds bankfull stage [23]. Given that inundation of the floodplain only occurs for a brief period of time, there is insufficient overbank flooding to hydrologically connect the lake to the main channel across the floodplain. Tres Chimbadas Lake drains to the main channel via a long, narrow (ca. 4 km long and 10 m wide) stream, and the lake appears to be relatively isolated in terms of fish movement. Thus, any MeHg that is bioaccumulated by fishes residing in the lake presumably originated from lake sediments. Although fishes from the lake generally had lower MeHg concentrations than fishes from the main channel, MeHg concentrations in some lake species were still higher than USEPA fish tissue criterion.

The Madre de Dios River and tributaries are currently unimpounded, but a large-scale hydroelectric project is being planned on the Inambari River. The 2200-megawatt dam would be the largest hydroelectric facility in Perú, flooding 46,000 hectares of land and causing extensive changes to the natural hydrologic regime. Because of widespread placer mining of gold in the Inambari River [10,23], there is large potential for extensive bacterial methylation of Hg within organic-rich sediments of the newly created reservoir. It is well documented that reservoir construction causes MeHg concentrations in fish tissue to rise [45-47]. Fishes inhabiting a reservoir on the Inambari River would be expected to have elevated MeHg concentrations.

Five out of the eight fish species surveyed from the main channel (63%) and two out of the five species from the lake (40%) had MeHg levels higher than USEPA fish tissue criterion. In the main channel, the fish species with the highest MeHg were migratory pimelodid catfish species of high market value. A recent study identifying levels of total mercury in human hair in the Madre de Dios Department found that mercury concentrations were significantly related to fish consumption level, gender, and location [mining zone versus city of Puerto Maldonado, 48]. Of all people surveyed that ate 12 or more fish meals per month, 18% had levels of mercury considered unhealthy (levels above 6.0 µg/g dry hair) by the World Health Organization [49]. Furthermore, 8% of men living in mining zones had mercury levels that are considered unhealthy. Indigenous communities in the Western Amazon have been documented to consume one to two meals of fish per day on average, and thus may be at risk of developing health problems related to MeHg exposure as a result of gold mining [50]. However, avoidance of fish consumption by indigenous people because of fear of Hg exposure can result in a switch to a western diet, associated with processed foods high in fat and low in protein [51]. Fish and fishing also have cultural significance, and avoidance of fish consumption has caused socio-cultural disruption in some indigenous Amerindian communities [52]. There are important implications for the endangered giant otter; at least one of the otter's preferred prey species, huasaco, had a high MeHg body burden. Five of the eight species from the main channel (63%) had MeHg concentrations greater than those known to cause neurological changes in the brain of the American mink, a species from the same family as giant otter [17]. Giant otters often use oxbow lakes as core areas within their home ranges, so it is possible that the high sediment organic matter and lower Hg levels found in fish in protected lakes where gold mining does not occur could partially protect them from greater MeHg exposure. More detailed analysis of bioaccumulation in aquatic food webs, human fish consumption surveys, and estimation of Hg levels in giant otters would greatly improve understanding of these risks.

ACKNOWLEDGEMENTS

We thank David Flores, Andrew Jackson, and Carmen Montaña for help in the field and Abir Biswas for technical assistance. Kurt Holle, Larissa Silva, Rob Williams, and other employees from the Frankfurt Zoological Society and Rainforest Expeditions assisted with local logistics. The Ese'eja native community of Infierno granted us access to the study areas. Hernan Ortega from the Museo de Historia Natural de Lima, Perú, provided aid in obtaining collection permits. Financial support came from University of Georgia Faculty Research Program (to CVF) and George and Carolyn Kelso (to KOW).

REFERENCES

1.	L. I. Sweet and J. T. Zelikoff, "Toxicology and Immunotoxicology of Mercury: A Comparative Review in Fish and Humans," Journal of Toxicology and Environmental Health, Part B Critical Reviews, Vol. 4, No. 2, 2001, pp. 161-205. doi:10.1080/109374001300339809

2.   H. M. Chan, A. M. Scheuhammer, A. Ferran, C. Loupelle, J. Holloway and S. Weech, "Impacts of Mercury on Freshwater Fish-Eating Wildlife and Humans," Human and Ecological Risk Assessment, Vol. 9, No. 4, 2003, pp. 867-883. doi:10.1080/713610013

3.   D. Borum, M. K. Manibusan, R. Schoeny and E. L. Winchester, "Water Quality Criterion for the Protection of Human Health: Methylmercury," Office of Science and Technology: Office of Water, US Environmental Protection Agency, Washington DC, 2001.

4.   J. Groenendijk and F. Hajek, "Giants of the Madre de Dios," Frankfurt Zoological Society, Lima, Peru, 2006.

5.   J. J. Swenson, C. E. Carter, J. C. Domec and C. I. Delgado, "Gold Mining in the Peruvian Amazon: Global Prices, Deforestation, and Mercury Imports," PLOS One, Vol. 6, No. 4, 2011, p. e18875. doi:10.1371/journal.pone.0018875

6.   CNN Money, "Commodities—Oil, Silver and Gold Prices," 2012.

7.   W. C. Pfeiffer and L. D. Lacerda, "Mercury Inputs into the Amazon Region, Brazil," Environmental Technology Letters, Vol. 9, No. 4, 1988, pp. 325-330. doi:10.1080/09593338809384573

8.   A. C. Gutleb, C. Schenck and E. Staib, "Giant Otter (Pteronura Brasiliensis) at Risk? Total Mercury and Methylmercury Levels in Fish and Otter Scats, Peru," Ambio, Vol. 26, No. 8, 1997, pp. 511-514.

9.   M. E. McClain and R. J. Naiman, "Andean Influences on the Biogeochemistry and Ecology of the Amazon River," BioScience, Vol. 58, No. 4, 2008, pp. 325-338. doi:10.1641/B580408

10.   M. Goulding, C. Cañas, R. Barthem, B. Forsberg and H. Ortega, "Amazon Headwaters: Rivers, Wildlife, and Conservation in Southeastern Peru," Asociación Para la Conservacióne la Cuenca Amazónica and Amazon Conservation Association, Lima, 2003.

11.   D. C. Depew, N. Basu, N. M. Burgess, L. M. Campbell, E. W. Devlin, P. E. Drevnick, C. R. Hammerschmidt, C. A. Murphy, M. B. Sandheinrich and J. G. Wiener, "Toxicity of Dietary Methylmercury to Fish: Derivation of Ecologically Meaningful Threshold Concentrations," Environmental Toxicology and Chemistry, Vol. 31, No. 7, 2012, pp. 1-12. doi:10.1002/etc.1859

12.   IUCN (International Union for Conservation of Nature), "IUCN Red List of Threatened Species, a Global Species Assessment," In:

J. E. M. Baillie, C. Hilton-Taylor and S. N. Stuart, Eds., Glandand Cambridge, 2004.

13. F. C. W. Rosas, J. Zuanon and S. K. Carter, "Feeding Ecology of the Giant Otter," Biotropica, Vol. 31, No. 3, 1999, pp. 502-506. doi:10.1111/j.1744-7429.1999.tb00393.x

14. C. Watanabe and H. Satoh, "Evolution of Our Understanding of Methylmercury as a Health Threat," Environmental Perspectives, Vol. 104, Suppl. 2, 1996, pp. 367- 379.

15. C. D. Wren, "Probable Case of Mercury Poisoning in a Wild Otter, Lutra canadensis, in Northwestern Ontario," Canadian Field Naturalist, Vol. 99, No. 1, 1985, pp. 112- 114.

16. A. C. Gutleb, A. Kranz, G. Nechay and A. Toman, "Heavy Metal Concentrations in Livers and Kidneys of the Otter (Lutra lutra) from Central Europe," Bulletin of Environmental Contamination and Toxicology, Vol. 60, No. 2, 1998, pp. 273-279.doi:10.1007/s001289900621

17. N. Basu, A. M. Scheuhammer, K. Rouvinen-Watt, N. Grochowina, K. Klenavic, R. D. Evans and H. M. Chan, "Methylmercury Impairs Components of the Cholinergic System in Captive Mink (Mustela vison)," Toxicological Sciences, Vol. 91, No. 1, 2006, pp. 202-209.doi:10.1093/toxsci/kfj121

18. C. C. Gilmour, E. A. Henry and R. Mitchell, "Sulfate Stimulation of Mercury Methylation in Fresh-Water Sediments," Environmental Science and Technology, Vol. 26, No. 11, 1992, pp. 2281-2287. doi:10.1021/es00035a029

19. G. Cabana and J. B. Rasmussen, "Modelling Food Chain Structure and Contaminant Bioaccumulation Using Stable Nitrogen Isotopes," Nature, Vol. 372, No. 6503, 1994, pp. 255-257. doi:10.1038/372255a0

20. K. A. Kidd, R. H. Hesslein, R. J. P. Fudge and K. A. Hallard, "The Influence of Trophic Level as Measured by ^{15}N on Mercury Concentrations in Freshwater Organisms," Water, Air, and Soil Pollution, Vol. 80, No. 1-4, 1995, pp. 1011-1015. doi:10.1007/BF01189756

21. D. M. Post, "Using Stable Isotopes to Estimate Trophic Position: Models, Methods, and Assumptions," Ecology, Vol. 83, No. 3, 2002, pp. 703-718. doi:10.1890/0012-9658(2002)083[0703:USITET]2.0.CO;2

22. M. A. Vanderklift and S. Ponsard, "Sources of Variation in Consumer-Diet [15]N Enrichment: A Meta-Analysis," Oecologia, Vol. 136, No. 2, 2003, pp. 169-182.doi:10.1007/s00442-003-1270-z

23. S. K. Hamilton, J. Kellndorfer, B. Lehner and M. Tobler, "Remote Sensing of Floodplain Geomorphology as a Surrogate for Biodiversity in a Tropical River System (Madre de Dios, Peru)," Geomorphology, Vol. 89, No. 1-2, 2007, pp. 23-38.doi:10.1016/j.geomorph.2006.07.024

24. M. E. Puhakka, R. Kalliola, M. Rajasilta and J. Salo, "River Types, Site Evolution and Successional Vegetation Patterns in Peruvian Amazonia," Journal of Biogeography, Vol. 19, No. 6, 1992, pp. 651-665. doi:10.2307/2845707

25. J. Groenendijk, F. Kajek, S. Isola and C. Schenk, "Giant Otter Project in Peru Field Trip and Activity Report— 2000," International Union for Conservation of Nature Otter Specialist Group Bulletin, Vol. 18, No. 2, 2001, pp. 76-85.

26. D. A. Arrington and K. O. Winemiller, "Preservation Effects on Stable Isotope Analysis of Fish Muscle," Transactions of the American Fisheries Society, Vol. 131, No. 2, 2002, pp. 337-342. doi:10.1577/1548-8659(2002)131<0337:PEOSIA>2.0.CO;2

27. C. W. Shade, "Automated Simultaneous Analysis of Monomethyl and Mercuric Hg in Biotic Samples by HgThiourea Complex Liquid Chromatography Following Acidic Thiourea Leaching," Environmental Science and Technology, Vol. 42, No. 17, 2008, pp. 6604-6610.doi:10.1021/es800187y

28. J. Olley, "Mercury in Fish and the News Media," Marine Pollution Bulletin, Vol. 4, No. 9, 1973, p. 143. doi:10.1016/0025-326X(73)90009-X

29. American Public Health Association, "Standard Methods for the Examination of Water and Waste-Water," 19th Edition, American Water Works Association, and Water Environment Federation, Washington, DC, 1992.

30. L. E. Fernandez and V. H. Gonzalez, "Niveles del Mercurio en Peces de Madre de Dios," Carnegie Institute for Science, Department of Global Ecology, Stanford University, Stanford, 2009.

31. N. Bloom, "On the Chemical Form of Mercury in Edible Fish and Marine Invertebrate Tissue," Canadian Journal of Fisheries and Aquatic Sciences, Vol. 49, No. 5, 1992, pp. 1010-1017. doi:10.1139/f92-113

32. A. C. Gutleb, A. Helsberg and C. Mitchell, "Heavy Metal Concentrations in Fish from a Pristine Rainforest Valley in Peru: A Baseline Study before the Start of Oil-Drilling Activities," Bulletin of Environmental Contamination and Toxicology, Vol. 69, No. 4, 2002, pp. 523-529. doi:10.1007/s00128-002-0093-7

33. C. A. Cooke, P. H. Balcom, H. Biester and A. P. Wolfe, "Over Three Millennia of Mercury Pollution in the Peruvian Andes," Proceedings of the National Academy of Sciences, Vol. 106, No. 22, 2009, pp. 8830-8834. doi:10.1073/pnas.0900517106

34. W. E. Brooks, E. Sandoval, M. A. Yepez and H. Howell, "Peru Mercury Inventory 2006," US Geological Survey, Open-File Report 2007-1252, 2007.http://pubs.usgs.gov/of/2007/1252/

35. Instituto Geológico Minero y Metalúrgico, Lima, 2009. http://www.ingemmet.gob.pe/form/Inicio.aspx

36. S. Ramamoorthy, S. Springthorpe and D. J. Kushner, "Competition for Mercury between River Sediment and Bacteria," Bulletin of Environmental Contamination & Toxicology, Vol. 17, No. 5, 1977, pp. 505-511. doi:10.1007/BF01685971

37. S. C. Choi and R. Bartha, "Environmental Factors Affecting Mercury Methylation in Estuarine Sediments," Bulletin of Environmental Contamination and Toxicology, Vol. 53, No. 6, 1994, pp. 805-812. doi:10.1007/BF00196208

38. R. F. C. Mantoura, A. Dickson and J. P. Riley, "The Complexation of Metals with Humic Materials in Natural Waters," Estuarine and Coastal Marine Science, Vol. 6, No. 4, 1978, pp. 387-408. doi:10.1016/0302-3524(78)90130-5

39. R. J. Breteler, I. Valiela and J. M. Teal, "Bioavailability of Mercury in Several North-Eastern U.S. Spartina Ecosystems," Estuarine, Coastal and Shelf Science, Vol. 12, No. 2, 1981, pp. 155-166. doi:10.1016/S0302-3524(81)80093-X

40. W. J. Langston, "Metals in Sediments and Benthic Organisms in the Mersey Estuary," Estuarine, Coastal and Shelf Science, Vol. 23, No. 2, 1986, pp. 239-261.doi:10.1016/0272-7714(86)90057-0

41. C. Y. Chen, M. Dionne, B. M. Mayes, D. M. Ward, S. Sturup and B. P. Jackson, "Mercury Bioavailability and Bioaccumulation in Estuarine Food Webs in the Gulf of Maine," Environmental Science and Technology, Vol. 43, No. 6, 2009, pp. 1804-1810. doi:10.1021/es8017122

42. K. O. Winemiller and D. B. Jepsen, "Effects of Seasonality and Fish Movement on Tropical River Food Webs," Journal of Fish Biology, Vol. 53, Suppl. A, 1998, pp. 267- 296. doi:10.1111/j.1095-8649.1998.tb01032.x

43. R. B. Barthem and M. Goulding, "The Catfish Connection: Ecology, Migration, and Conservation of Amazon Predators," Columbia University Press, New York, 1997.

44. M. J. de Jesús and C. C. Kohler, "The Commercial Fishery of the Peruvian Amazon," Fisheries, Vol. 29, No. 4, 2004, 10-16. doi:10.1577/1548-8446(2004)29[10:TCFOTP]2.0.CO;2

45. A. T. Jackson, "The Mercury Problem in Recently Formed Reservoirs of Northern Manitoba (Canada): Effects of Impoundment and Other Factors on the Production of Methyl Mercury by Microorganisms in Sediments," Canadian Journal of Aquatic Sciences, Vol. 45, No. 1, 1988, pp. 97-121. doi:10.1139/f88-012

46. R. Schetagne, J. F. Doyon and J. J. Fournier, "Export of Mercury Downstream from Reservoirs," Science of the Total Environment," Vol. 260, No. 1-3, 2000, pp. 135- 145.doi:10.1016/S0048-9697(00)00557-X

47. L. D. Hylander, J. Gröhn, M. Tropp, A. Vikström, H. Wolpher, E. de Castro e Silva, M. Meili and L. J. Oliveria, "Fish Mercury Increase in Lago Manso, a New Hydroelectric Reservoir in Tropical Brazil," Journal of Environmental Management, Vol. 81, No. 2, 2006, pp. 155- 166. doi:10.1016/j.jenvman.2005.09.025

48. K. Ashe, "Elevated Mercury Concentrations in Humans of Madre de Dios, Peru," PLOS One, Vol. 7, No. 3, 2012, p. e33305. doi:10.1371/journal.pone.0033305

49. World Health Organization, "Environmental Health Criteria. I. Mercury," World Health Organization, Geneva, 1976.

50. A. A. P. Boischio and D. Henshel, "Fish Consumption, Fish Lore, and Mercury Pollution—Risk Communication for the Madeira

River People," Environmental Research Section A, Vol. 84, No. 2, 2000, pp. 108-126. doi:10.1006/enrs.2000.4035

51. B. Wheatley and S. Paradis, "Balancing Human Exposure, Risk and Reality: Questions Raised by the Canadian Aboriginal Methylmercury Program," Neurotoxicology, Vol. 17, No. 1, 1996, pp. 241-249.

52. B. Wheatley and M. A. Wheatley, "Methylmercury and the Health of Indigenous Peoples: A Risk Management Challenge for Physical and Social Sciences and for Public Health Policy," Science of the Total Environment, Vol. 259, No. 1-3, 2000, pp. 23-29.doi:10.1016/S0048-9697 (00)00546-5.

Analysis of Radium Isotopes in Surface Waters nearby a Phosphate Mining with NORM at Santa Quitéria, Brazil

Wagner de S. Pereira[1, 2], Alphonse Kelecom[2],
and Juliana R. de S. Pereira[3]

[1]Multidisciplinary Group of Radioprotection (GMR), Serviço de Radioproteção, Unidade de Tratamento de Minério, Indústrias Nucleares do Brasil, Caldas, Brazil

[2]Laboratory of Radiobiology and Radiometry Pedro Lopes dos Santos (LARARA-PLS), Group of Environmental Themes Assessment (GETA), Universidade Federal Fluminense (UFF), Niterói, Brazil

[3]Interdisciplinary UnderGraduation in Science and Technology, Universidade Federal de Alfenas, Poços de Caldas, Brazil

ABSTRACT

The radium isotopes ^{226}Ra and ^{228}Ra were analyzed in surface water at six points in the neighborhood of a mine of phosphate, associated

with uranium, in the region of Santa Quitéria, state of Ceará, Brazil. Water samples were collected during twenty months, filtered and the concentrations of activity determined in the soluble and particulate phases. The results were analyzed using the Principal Component Analysis (PCA) for ordination of environmental data, and also by ANOVA, Tukey and Z tests to compare sets of data considering the radionuclides, the two analyzed phases and the six collecting points. The PCA identified four groups that included all collecting points, using aggregation features such as radionuclide and analyzed phase. The first group is composed by the samples of ^{226}Ra in the soluble phase; the second group by samples of ^{226}Ra in the particulate phase; the third one by ^{228}Ra in the soluble phase, and finally, the fourth group by ^{228}Ra in the particulate phase. This last group has two discrepant points (01 and 06). Statistical analysis identified differences between the concentrations of activity of radionuclides (^{228}Ra higher than ^{226}Ra) and in analyzed phases (soluble phase higher than the particulate one) but showed no differences between sampled points.

INTRODUCTION

Phosphate mining and processing can cause significant radiological impacts due to the amount of radionuclides present in the ore [1]. Phosphogypsum e.g., a by-product of phosphate mining, is contaminated by heavy metals and radionuclides, especially ^{226}Ra [2]. Mining of phosphate with associated Naturally Occurring Radioactive Materials must thus be considered as a NORM activity.

There has been an increased awareness of the radiological impacts of NORM non-nuclear mining pointing that this activity may cause radioactive contamination due to the by-products, wastes and to the installations themselves [3]. In this respect, the environmental impact was analyzed in three practices related to phosphate production: mines, phosphate fertilizers factory and phosphate export platforms. Air particulates, soil, water (lake, river and sea water), biota and plant samples were collected and analyzed. An increase of natural radionuclides in the surroundings of the three enterprises was observed, with fallout being the principal contamination way [4].

Accordingly, the phosphate industry has been recently included within the European regulatory scope. ^{226}Ra was recognized as the major contaminant. It is found mainly in the processing waters, and the major releasing way is via the liquid effluents [5].

Located in the central-north region of the state of Ceará, Brazil, the "Santa Quitéria" Unit is a phosphate associated with uranium mining installation in predevelopment stage that belongs to the "Indústrias Nucleares do Brasil" (INB). The deposit has recoverable reserves of about nine million tons of P_2O_5 and 79,500 tons of U_3O_8 [6]. The unit is under the influence of the "Bsh" semiarid climate. The raining season extends from January to May, with sporadic precipitations in June and July. The annual rainfall varies from 550 mm to 960 mm [6]. The rainiest month is March, with rainfall indices varying from 115 mm to 230 mm [6]. Ecologically, the area is characterized as a tropical steppe known as caatinga (savanna) with tropical forests and human occupation areas.

According to Brazilian norms, the unit is classified as a NORM installation [7]. The development of a unit with this classification demands a radiological environmental monitoring program [7-10]. In this context, monitoring means a systematic and planning process of measuring radiation fields, radioactivity and other environmental parameters, including the interpretation of these measurements, in order to characterize, evaluate and control public exposure, especially the critical group, the most exposed to radiation resulting from practice [7-10].

In terms of radioecology, it is necessary to evaluate the behavior of the radionuclides and of their phases in the environment, and how this behavior is able to alter the concentrations of activity of these parameters.

In terms of environmental radiation protection, it is necessary to analyze the possible changes in behavior and in concentrations of activities along the ways of exposition of population and biota in order to evaluate the radiological environmental impact of the project.

A number of radionuclides must be analyzed aiming a comprehensive understanding of their behavior and dispersion in the region. A model for assessment of environmental radiological impact should also be proposed in order to estimate the exposition before the operation and to allow the licensing based in terms of increased dose caused by the practice, as determined by Brazilian laws [2-5].

In Spain, the radiological impact of phosphate mining with NORM decreased significantly with the regulation of the practice and with the separation of the regulation on radiological aspects and chemical aspects that coexist in this kind of practice [11]. The authors also point the extensive regulation of the European Union countries in both aspects of the impact of mining [11].

We here report on the concentrations of activity of the radium isotopes [226]Ra and [228]Ra in surface water measured at six monitoring points in the vicinity of the Santa Quitéria mine. The values were submitted to usual statistical treatment such as the ANalysis Of VAriance (ANOVA), and the Tukey and Z tests [12-15], but also to a data ordination technique used in multivariate statistics known as the Principal Component Analysis (PCA), as an auxiliary tool for the interpretation of measured data [16-18]. The use of univariate and multivariate analyses in the pre-operational environmental monitoring aimed the licensing of the phosphate mine with NORM characteristics, using radium isotopes as a case study.

METHODOLOGY

Sampling Area

The Santa Quitéria Unit is located in the municipality of Santa Quitéria, state of Ceará, 212 Km South from the state capital Fortaleza, in northeast of Brazil (Figure 1).

Sample Collection and Preparation

Environmental water samples (one liter each) were collected monthly during twenty months: in January 2006, and then continuously from June 2006 until December 2007, at six points around the unit, as shown in Table 1. All samples were sent to the Federal University of Ceará where they were filtered through a cellulose acetate filter of porosity 0.45 μm. The fraction that passes through the filter was considered as the soluble phase and material retained on the filter was considered as the particulate phase. After filtration, samples were acidified with 1 ml conc. nitric acid per liter and finally sent to the laboratory of environmental analysis of the "Indústrias Nucleares do Brasil" (INB), at the Ore Treatment Unit (UTM) situated at Poços de Caldas, state of Minas Gerais, Brazil, for radionuclides determinations.

Figure 1: Location of Santa Quitéria municipality in the state of Ceará, Brazil (adapted from Wikipedia).

Table 1: Geographical locations of water collecting points in "UTM" coordinates

Points	UTM E	UTM N
01	0408575	9495305
02	0409766	9496141
03	0415241	9495593
04	0413784	9493155
05	0410362	9494800
06	0411850	9494292

Radionuclides Determination

The determined radionuclides were ^{226}Ra and ^{228}Ra. Radium has been chemically separated by co-precipitation from the other radionuclides present in the samples. ^{226}Ra has been determined by gross alpha counting and ^{228}Ra has been determined by gross beta counting, as described elsewhere [19].

Sampling Design and Data Organization

Data were separated by month of collection, collecting point (numbered 01, 02, 03, 04, 05 and 06), radionuclide (^{226}Ra or ^{228}Ra), and sample phase (particulate or soluble phase). Data were organized in a matrix having 24 columns referring to the six collecting points split for each radionuclide and each sample phase, and 20 lines referring to the 20 months of collection. Collected data sum a total of 480 values to be analyzed. Then, the data were grouped in several ways: by radionuclides, regardless of the phase and collecting point, by phase regardless of the radionuclide and the collection point, and finally by collecting point regardless of phase and radionuclide. Thus three factors were analyzed: radionuclide, sample phase and collecting point.

One Variable Statistic Analysis

The statistical tests used in univariable statistics require adjustments to the Gaussian distribution [12-15]. For this, the adjusting Anderson-Darling test was performed, using the statistical package Minitab® version 16.

An ANOVA was performed to verify the existence of differences in radium concentrations of activity between points, radionuclides and their fractions. This analysis was carried out with the statistical package Excel® version 2010 for Windows® environment. Existing differences, the Tukey test was performed to "group" points, radionuclides and phases with the same concentration of activity. The statistical package Minitab® version 16 was used for this analysis. Tested hypotheses by ANOVA were:

- H0 there are no differences between analyzed means;
- H1 there is at least one different mean.

Another ANOVA was executed between points grouping all results of each point regardless the radionuclide or phase. This analysis was performed with the statistical package Excel® version 2010 for Windows® environment. Existing differences the Tukey test was applied to "group" the points with the same concentration of activity.

Four Z tests were made: one to compare the concentrations of activity in the phases (soluble and particulate) for ^{226}Ra isotope and a second identical for isotope ^{228}Ra. The third one was carried out between the phases, regardless the radionuclide (i.e. data for the soluble phases for ^{228}Ra and ^{226}Ra versus data for the particulate phases for both isotopes). Finally, the fourth Z test was performed with the radionuclides regardless of the phases (i.e. data on ^{226}Ra in the soluble and particulate phases versus data on ^{228}Ra in the same phases). These analyzes were done with the statistical package Excel® version 2010 for Windows® environment. Tested hypotheses were:

- H0 there are no differences between means;
- H1 there are differences between means.

Multi Variable Statistic Analysis

Principal Component Analysis

Principal Component Analysis (PCA) is a technique for modeling co-variances, which was introduced in 1901 by Pearson [16]. Although multivariate techniques for environmental data assessment are a need, since the world is composed by multi-factorial inter-related parameters, Valentin reported that the first application of PCA in ecology occurred only in 1954 [17].

Nowadays PCA is the most used ordination technique for ecological analysis. Environmental data are ordered in one or two axis. The parameters are established by a correlation similarity of variance-covariance matrix [17]. PCA uses this matrix to produce a set of orthogonal axes ordered from highest to lowest values of a parameter (factor) according to its contribution to the total variance of data. The result is a reduced system of coordinates in which both the position of data in relation to the axes and the relationship between data provide information on the similarities of environmental data [16-18].

RESULTS

Frequency Distribution

Data were analyzed by the Anderson-Darling test for fitness to normal distribution. Concentrations of [226]Ra in the particulate phase gave a test parameter of 15.027 and [226]Ra in the soluble phase a value of 28.064. For [228]Ra, the value in the particulate phase was 8.231 and in the soluble phase 30.422. For all these tests the critical value was less than 0.05. Thus, the four distributions were considered log-normal distributed and therefore a normalizing process was required in the form y=in(x+1), as recommended by Ceteno [12].

Radium Activity Concentrations

Concentrations of activity of [226]Ra and [228]Ra in water samples, collected during twenty months, were obtained according to described methodology [19]. Table 2 shows the averages of the concentrations of activity and the number of data analyzed for both radionuclides, organized by collecting point and sample phase (before logarithmic transformation). These results can be seen in a graphical way in Figure 2.

One Variable Analysis

The ANOVA was realized between points, radionuclides and their phases after logarithmic normalizing transformation in the form [y=in(x+1)] of the values reported in Table 2. In this case, F_{cal} (value calculated by the test) is equal to 2.28 being higher than F_{cri} (value if accepted the H_1 hypothesis) that is equal to 1.55, with an associated P (statistically significant result) lower than 0.001. This analysis showed differences in the concentrations of activity between the points, radionuclides and phases.

After performing the ANOVA on normalized data, a Tukey test was carried out to detect groups of data with similar average concentrations of activity. Three groups were observed (Table 3).

The first group with the highest mean concentrations of activity appears under lable "A" alone (Table 3). This group has only one representative, i.e. ^{228}Ra in the soluble phase at point 06.

The second group with intermediate concentrations of activity is composed by the 11 factors labeled "A" and "B" simultaneously (Table 3). This group is formed by representatives of ^{226}Ra and ^{228}Ra in both soluble and particulate phases, that is ^{228}Ra in the soluble phase at points 02, 03, 04 and 05; ^{228}Ra in the particulate phase at points 02, 03, 05 and 06; ^{226}Ra in the soluble phase at points 01 and 03, and ^{226}Ra in the particulate phase at point 01.

Finally, the third group with the lowest concentrations of activity is composed by the 12 factors labeled "B" only (Table 3). This group, likewise the second one, is formed by representatives of ^{226}Ra and ^{228}Ra in both soluble and particulate phases. This group contains ^{228}Ra in the soluble phase at point 01; ^{228}Ra in the particulate phase at points 01 and 04; ^{226}Ra in the soluble phase at points 02, 04, 05 and 06, and ^{226}Ra in the particulate phase at points 02, 03, 04, 05 and 06.

Another ANOVA, on normalized data, was performed between collecting points regardless of the phases and radionuclides. This analysis showed the absence of statistical differences between collecting points as F_{cal} (0.96) was lower than F_{cri} (2.23), with P = 0.45. Thus, all collecting points are considered to have identical means of concentration of activity when the factors radionuclide and phase are not considered.

Table 2: Averages over 20 months of the concentrations of activity of ^{226}Ra and ^{228}Ra in the soluble and particulate phases at the 6 collecting points (Bq·l^{-1})

Point	^{226}Ra		^{228}Ra		N
	Soluble	Particulate	Soluble	Particulate	
01	0.04139	0.01221	0.00889	0.00419	20
02	0.00806	0.00424	0.01386	0.01400	20
03	0.03057	0.00489	0.01974	0.02551	20
04	0.00448	0.00346	0.06893	0.00715	20
05	0.00669	0.00377	0.07362	0.01894	20
06	0.00963	0.00318	0.09743	0.03756	20

Table 3: Tukey test for grouping the averages

Factor radionuclide-phase-point		N	Average	grouping	
Ra-228 soluble	06	20	0.07925	A	
Ra-228 soluble	05	20	0.05880	A	B
Ra-228 soluble	04	20	0.05397	A	B
Ra-226 soluble	01	20	0.03591	A	B
Ra-228 particulate	06	20	0.03571	A	B
Ra-226 soluble	03	20	0.03004	A	B
Ra-228 particulate	03	20	0.02463	A	B
Ra-228 soluble	03	20	0.01941	A	B
Ra-228 particulate	05	20	0.01840	A	B
Ra-228 particulate	02	20	0.01377	A	B
Ra-228 soluble	02	20	0.01331	A	B
Ra-226 particulate	01	20	0.01189	A	B
Ra-226 soluble	06	20	0.00957		B
Ra-228 soluble	01	20	0.00851		B
Ra-226 soluble	02	20	0.00796		B
Ra-226 soluble	05	20	0.00666		B
Ra-228 particulate	04	20	0.00636		B
Ra-226 particulate	03	20	0.00485		B
Ra-226 soluble	04	20	0.00447		B
Ra-226 particulate	02	20	0.00423		B
Ra-228 particulate	01	20	0.00377		B
Ra-226 particulate	05	20	0.00375		B
Ra-226 particulate	04	20	0.00344		B
Ra-226 particulate	06	20	0.00317		B

Comparing among themselves the values of the soluble and particulate phases for each isotope (^{226}Ra or ^{228}Ra), it appears that the mean concentrations of activity of ^{226}Ra in the phases are considered statistically different ($Z_{cal} = 2.75 > Z_{cri} = 1.64$, with P < 0.01), with the values in the soluble phase higher than in the particulate one. The same behavior was observed for ^{228}Ra with values of the soluble phase again higher than of the particulate one ($Z_{cal} = 2.08 > Z_{cri} = 1.64$, P < 0.01).

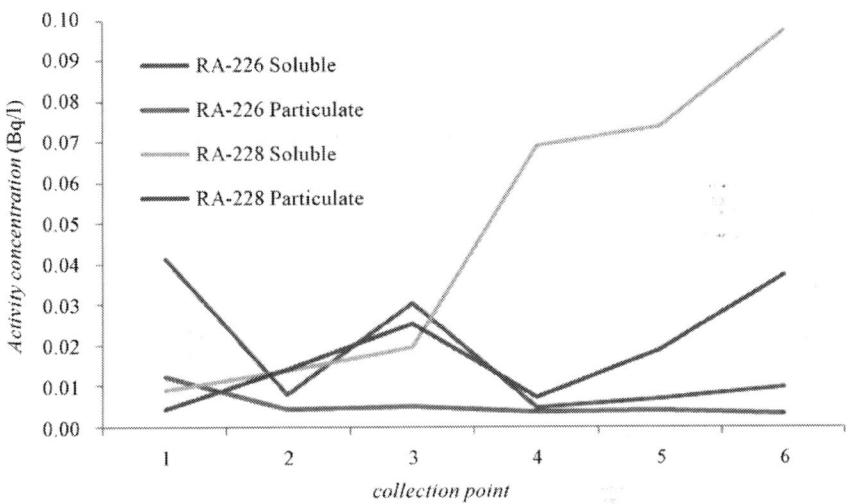

Figure 2: Activity concentration average values of ^{226}Ra and ^{228}Ra in particulate and soluble phases at the 6 collecting points (Bq·l^{-1}).

When the values of the phases are analyzed regardless of the radionuclides and of the collecting points, the values for the soluble phase were considered statistically higher than those of the particulate phase ($Z_{cal} = 6.97 > Z_{cri} = 1.64$, P < 0.01).

Now comparing the radionuclides regardless of the phases and of the collecting points, the concentrations of activity for ^{228}Ra were considered higher than those of ^{226}Ra ($Z_{cal} = 2.85 > Z_{cri} = 1.64$, with P < 0.01).

Ordination of Data

The PCA analysis results are shown in Figure 3. PCA identified four groups of samples. The results of ^{228}Ra in the soluble phase are grouped in the negative parts of axis 1 (Factor 1) and axis 2 (Factor 2).

Results of ^{226}Ra in the particulate phase are grouped in the positive part of axis 1 (Factor 1) and in the negative part of axis 2.

Results of ^{226}Ra in the soluble phase are grouped along the positive part of axis 1 (Factor 1) and near the origin of axis 2 (Factor 2).

Finally, results of ^{228}Ra in the particulate phase are grouped in the positive parts of axis 1 (Factor 1) and axis 2 (Factor 2). This group shows two discrepant points (01 and 06), due to their variance most related to the group composed by ^{228}Ra in the soluble phase, mainly point 06.

DISCUSSION AND CONCLUSIONS

The statistical analyses allow commentaries. When the radionuclides are analyzed regardless of the other variables, it may be concluded that there are differences between the concentrations of activity of considered radionuclides, with ^{228}Ra showing higher concentrations of activity than ^{226}Ra.

When the phases are analyzed regardless of the other variables, differences are observed between them, with the soluble phase showing higher concentrations of activity than the particulate one.

Concerning the collecting points, another scenario is observed. Thus, when the points are analyzed regardless of the other variables, no differences were observed between them. All the points were considered to have identical means.

Thus, the univariate analysis (ANOVA) allowed establishing that the environmental variables "radionuclide" and "phase" have different behaviors, but the variable "collecting point" showed no differences between the points.

The situation is more complicated when the collecting points are analyzed taking into consideration the variables radionuclide and phase. Three distinct groups appear, with high, intermediate or low means.

The highest average appears for ^{228}Ra in the soluble phase at point 06, forming a group of a single element. The second group with intermediate averages is heterogeneous being composed by ^{228}Ra, in both phases at points 02, 03 and 05, plus ^{228}Ra in the soluble phase at point 04 and in the particulate phase at point 06, and also by ^{226}Ra in both phases at point 01 and in the soluble phase at point 03. The last group, with the lowest concentrations of activity, is composed by ^{228}Ra in both phases at point 01 and ^{228}Ra in the particulate fraction at point 04. This group also contains ^{226}Ra in both phases at points 02, 04, 05 and 06 together with ^{226}Ra in the particulate fraction at point 03.

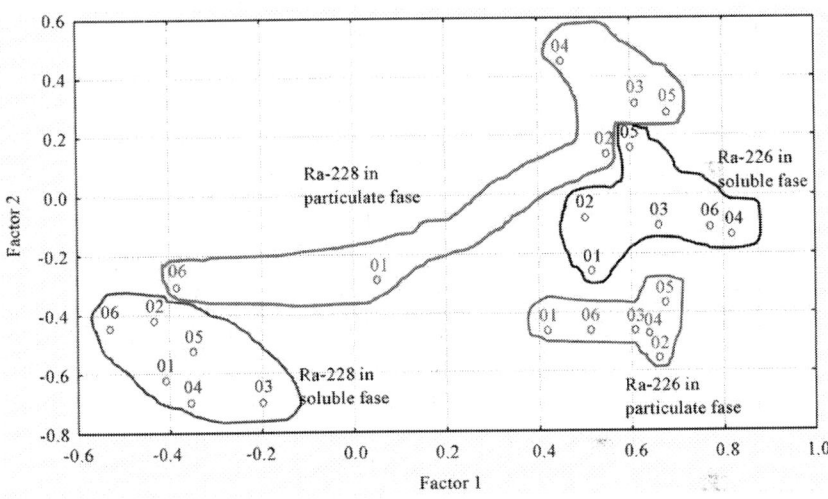

Figure 3: Ordination of collecting points as a function of radionuclides and sample phase.

The Tukey test detected the trend of Ra-228 to have higher concentrations than Ra-226, and this was corroborated by the Z test. Thus, Table 3 reports Ra-228 as the only representative of the group of highest activity (A); it is also predominant in the group of intermediate activity AB (8 hits in 11, 73%) but uncommon in the group of lowest activity B (3 in 12, 25%).

In relation to the phase of the radionuclide, values showed no trend. On the contrary, the univariate analysis (Tukey test) identified differences between the collecting points, forming groups of similar activity concentration.

Complementary information was achieved using multivariated analysis. Thus, the PCA technique was used here as a tool to evaluate the possibility of grouping the collecting points depending of the environmental variables radionuclide and phase. Indeed, PCA enabled to ordinate the six collecting points in four groups (Figure 3): one associated the six points with ^{226}Ra in the soluble phase; a second the six points with ^{226}Ra in the particulate phase and a third the six points with ^{228}Ra in soluble phase. Finally the fourth group associated the points with ^{228}Ra in the particulate phase. In the latter group, two points (01 and 06) are quite distant producing an irregular group, different from the other three which are nicely homogeneous.

Thus on the whole, PCA was a good method for the ordination of data from monitoring points using concentration of activity of radionuclides and sample phases as parameters.

The combination of univariate and multivariate statistical analyses enabled to assemble a more comprehensive analysis of pre-operational environmental monitoring at the Santa Quitéria phosphate with NORM mine, affording complementary information that only one class of statistical analysis cannot furnish.

Data from this study represent the values of activity concentration in the region before the beginning of the operation of the mine (background). The mine operation inevitably will affect the overall picture presented here and such changes will need to be analyzed in order to assess major concerns relative to radioecology and environmental radioprotection.

REFERENCES

1. E. M. Ashraf, H. A. Khater and A. L. Sewaidan, "Radiation Exposure Due to Agricultural Uses of Phosphate Fertilizers," Radiation Measurements, Vol. 43, No. 8, 2008, pp. 1402-1407. http://dx.doi.org/10.1016/j.radmeas.2008.04.084

2. P. M. Rutherford, M. J. Dudas and J. M. Arocena, "Radio-Activity and Elemental Composition of Phosphogypsum Produced from Three Phosphate Rock Sources," Waste Management & Research, Vol. 13, No. 5, 1995, pp. 407-423.

3.   H. Beddow, S. Black and D. Read, "Naturally Occurring Radioactive Material (NORM) from a Former Phosphoric Acid Processing Plant," Journal of Environmental Radioactivity, Vol. 86, No. 3, 2006, pp. 289-312. http://dx.doi.org/10.1016/j.jenvrad.2005.09.006

4.   I. Othman and M. S. Al-Masri, "Impact of Phosphate Industry on the Environment: A Case Study," Applied Radiation and Isotopes, Vol. 65, No. 1, 2007, pp. 131- 141.http://dx.doi.org/10.1016/j.apradiso.2006.06.014

5.   N. Casacuberta, P. Masqué and J. Garcia-Orellana, "Fluxes of ^{238}U Decay Series Radionuclides in a Dicalcium Phosphate Industrial Plant," Journal of Hazardous Materials, Vol. 190, No. 1-3, 2011, pp. 245-252. http://dx.doi.org/10.1016/j.jhazmat.2011.03.035

6.   J. R. Silva, "Caracterização Hidrogeológica da Jazida de Itataia," M.Sc. Dissertation, Universidade Federal do Ceará, Ceará, 2003, p. 126.

7.   National Commission for Nuclear Energy—CNEN, Standard CNEN-NN-4.01, "Requisitos de Segurança e Proteção Radiológica para Instalações Mínero-Industriais," 2005, p. 19.

8.   National Commission for Nuclear Energy—CNEN, Standard CNEN-NN-3.01, "Diretrizes Básicas de Proteção Radiológica," 2005a, p. 34.

9.   National Commission for Nuclear Energy—CNEN, "Standard Posição Regulatória 3.01/008:2011 Programa de Monitoração Radiológica Ambiental," 2011, p. 5.

10. National Commission for Nuclear Energy—CNEN, "Standard Posição Regulatória 3.01/009:2011 Modelo para Elaboração de Relatórios de Programa de Monitoração Radiológica Ambiental," 2011, p. 5.

11. M. García-Talavera, J. L. M. Matarranz, R. Salas and L. Ramos, "A Regulatory Perspective on the Radiological Impact of NORM Industries: The Case of the Spanish Phosphate Industry," Journal of Environmental Radioactivity, Vol. 102, No. 1, 2011, pp. 1-7. http://dx.doi.org/10.1016/j.jenvrad.2010.08.007

12. A. J. Ceteno, "Curso de Estatística Aplicada à Biologia," Universidade Federal de Goiás, Goiânia, 1999, p. 188.

13. H. G. Arango, "Bioestatística: Teórica e Computacional Com Bancos de Dados Reais em Disco," 3rd Edition, Guanabara & Koogan, 2009, p. 438.

14. M. A. Schork and R. D. Remington, "Statistics with Applications to the Biological and Health Science," 3rd Edition, Prentice Hall, Upper Saddle River, 2000, p. 478.

15. B. Rosner, "Fundamentals of Biostatistics," 5th Edition, Duxbury Thomson Learning, Stamford, 2000, p. 792.

16. D. F. Ferreira, "Estatística Multivariada," Editora da UFLA, Lavras, 2008, p. 650 p.

17. J. L. Valentin, "Ecologia Numérica. Uma Introdução à Análise Multivariada de Dados Ecológicos," Interciência, Rio de Janeiro, 2000, p. 117.

18. N. J. Gotelli and A. M. Ellison, "A Primer of Ecological Statistic," Sinauer Associates, Sunderland, 2004, p. 510.

19. J. M. Godoy, D. C. Lauria, M. L. D. P. Godoy and R. P. Cunha, "Development of a Sequential Method for Determination of ^{238}U, ^{234}U, ^{232}Th, ^{230}Th, ^{228}Th, ^{228}Ra, ^{226}Ra and ^{210}Pb in Environmental Samples," Journal of RadioAnalytical Nuclear Chemistry, Vol. 182, No. 1, 1994, pp. 165-169. http://dx.doi.org/10.1007/BF02047980

Chapter 10

A Survey of Experience Gained from the Treatment of Coal Mine Wastewater

Estêvão A. Pondja Jr[1, 2], Kenneth M. Persson[1],
and Nelson P. Matsinhe[2]

[1]Department of Building and Environmental Technology, Lund University, Lund, Sweden

[2]Department of Chemical Engineering, Eduardo Mondlane University, Maputo, Mozambique

ABSTRACT

During coal mining, water resources may be polluted by acid mine drainage (AMD) if appropriate measures are not taken. AMD releases metals to the environment, which can be harmful to aquatic species and reduce biodiversity. There is a great deal of information available in the literature on the generation and treatment of AMD and this paper tries to summarize some of them in order to facilitate the choice of the most appropriate method for AMD treatment at a specific mining site. The objective of this study was to identify and describe different methods of treating polluted water from coal mining, and to discuss the choice of suitable methods at specific mining sites. Both active and passive methods of AMD treatment are discussed in order to provide a general picture of the measures that have been taken around the world by coal mining companies. From this study, we were able to conclude that in order to choose the appropriate method for a specific mining site it is necessary to analyze the chemistry of the acid water and the flow rate from that site. The cost, implement ability and effectiveness of the method should also be considered. Minimizing the amount of drainage water generated is naturally the first choice of management strategy and the containment of the AMD is the second choice. The third alternative is the treatment of the wastewater.

INTRODUCTION

Coal mining plays an important role worldwide in both the energy and metallurgical industries. Thermal coal for the production of electricity and coking coal for steel production are the main products of the coal mining industry. Coal mining activities also produce solid waste, and air and water pollutants. Acid mine drainage (AMD) is the main environmental problem caused by mining activities.

Acidic leachate can occur naturally, due to the weathering of minerals containing sulfides, leading to the oxidation of elemental sulfur, but the greatest sources of acidic wastewater arise from anthropogenic activities such as mining [1]. Mining accelerates the process of weathering of reactive sulfide by increasing the available surface area of reactive components allowing enormous amounts of material containing sulfides to be exposed to air and water [1]. The most dominant sulfide mineral in many ore deposits is pyrite, and this plays a key role in the generation of AMD [2]. However, other sulfide minerals are also present, and their oxidation also affects mine water chemistry. Pyrite, pyrrhotite, marcasite and mackinaw wite are the most reactive sulfides, and their oxidation results in water with a low pH [2]. Sulfide minerals are formed in the absence of oxygen in ore mineral deposits, i.e., they are formed under reducing conditions and will become unstable when exposed to oxygen, for example, in mining water, and during excavation, mineral processing and other activities that involve the removal of mineral-containing material [1] .

The generation of AMD can be explained by Equations (1)-(5). Pyrite can react directly with oxygen forming an acidic solution Equation (1), and this reaction can take place in the presence or absence of microorganisms. Ferric iron (Fe^{3+}) dissolved in water can oxidize pyrite, Equation (2), and the ferric iron is replenished by the oxidation of ferrous iron in the presence of aerobic bacteria, which catalyze the reaction in Equation (3). Oxidation and hydrolysis of ferrous iron (Fe^{2+}) under slightly acidic to alkaline conditions lead to the formation of an insoluble hydroxide, Equation (4). When reactions (1) and (4) take place at a pH above 4.5, Equation (5) results, and the acidity is doubled compared to reaction 1.

For example:

$$FeS_2 + \frac{7}{2}O_2 + H_2O = Fe^{2+} + 2SO_4^{2-} + 2H^+ \tag{1}$$

$$FeS_2 + 14Fe^{3+} + 8H_2O = 12Fe^{2+} + 2SO_4^{2-} + 16H^+ \tag{2}$$

$$Fe^{2+} + \frac{1}{4}O_2 + H^+ = Fe^{3+} + \frac{1}{4}H_2O \tag{3}$$

$$Fe^{2+} + \frac{1}{4}O_2 + 2\frac{1}{2}H_2O = Fe(OH)_3 + 4H^+ \tag{4}$$

$$FeS_2 + \frac{15}{4}O_2 + \frac{7}{2}H_2O = Fe(OH)_3 + 2SO_4^{2-} + 4H^+ \tag{5}$$

Sources of AMD

The main sources of AMD are ore and coal stockpiles, tailing storage facilities, waste rock piles, leach piles, mine adits, and pit walls, shafts and floors [2]. Rocks containing sulfides are considered to be one of the major sources of AMD, and their management is thus very important [3]. The composition of AMD depends on the mineralogy of the local rocks, and water and oxygen availability, and thus every mine is unique with regard to its potential to generate AMD [3].

Acid-Buffering Reactions

Rocks normally contain alkaline materials such as carbonates (calcite and dolomite), silicates and hydroxide, which can neutralize AMD [4]. Silicates constitute the largest reservoir of buffering capacity on Earth, but a wide range of calcites also occur, and these are considered to be the most important neutralizing agent due to their rapid reaction rate compared with silicates [2] . When AMD interacts with alkaline material in the rocks, some of the acidity is neutralized, which means that not all the leachate from waste or stock piles at a mining site will generate an acid solution. To determine whether a certain waste containing sulfide can generate acid water it is necessary to perform static and kinetic tests. Static test determines balance between neutralizing potential and acid potential of mine waste while kinetic test provide information about leachate quality and rate [5].

Impact of AMD on the Environment

AMD can have several effects on the environment, the main ones being the release of metals into waterways causing the death of fish and other aquatic species. Fish may also become contaminated by eating contaminated sediment and food, due to the high content of metals in the water [6]. One of the main products of pyrite oxidation is iron hydroxide ($Fe(OH)_3$), which precipitates in streams giving them a red/orange color (Figure 1). It can also cover the surface of sediments and stream beds, contributing to the destruction of habitats [6].

Wastewater Generation from Coal Mining

Solid, liquid and gaseous effluents are produced during mining, and mining companies should take measures to minimize or eliminate these effluents in order to achieve sustainable production. The subject of this study is the acid water produced by mining. Wastewater resulting from coal mining can be divided into mine water, process wastewater, domestic wastewater and storm water [7]. Mine water can be defined as the ground or surface water at a mining site [2]. Process water can be divided into liquid effluent and tailings. Wastewater resulting from machinery, the washing of trucks and working areas, and pipe leakage are considered to be liquid effluent. This type of wastewater contains a high level of non-filterable residue Waste resulting from the coal washing process is called slurry tailings and is a potential source of acid water [7]. AMD can be generated when storm water comes into contact with the surface of sulfide-containing minerals (e.g. pits walls) or overburden. Precipitation can seep through waste piles resulting in groundwater contamination. Domestic wastewater arises from offices around the mining area. If fine particles of coal are not removed from employees' clothing and bodies after mining, they may find their way into domestic wastewater as a result of washing.

Control of AMD

AMD can be controlled using 3 different techniques: prevention, containment and remediation (treatment).The aim of prevention is to completely avoid the generation of acid water by avoiding contact between sulfide-con- taining minerals and water/oxygen The common methods used are the isolation of metallic sulfide, oxygen exclusion using wet and dry covers, and alkaline additives [8] . The aim of containment is to avoid flows of AMD to the environment. Some of the methods used are impoundment of AMD, alkaline-permeable barriers, and the disposal of tailings in impermeable cells [8]. The aim of remediation is to increase the pH and reduce the concentrations of pollutants such as metals, solids and salts present in AMD, to avoid contamination of surface water and groundwater [8]. Remediation methods can be divided into active and passive treatment.

Other strategies can be used to reduce the amount of water requiring treatment, such as the construction of upstream dams to intercept and divert surface water, the avoidance of seepage of rain water to contaminated areas, maximization of the reuse or recycling of water, separation of water with different qualities, the avoidance of infiltration of contaminated water into the groundwater, and appropriate management of waste containing sulfides [3].

METHODS OF TREATING COAL MINING WASTEWATER

In cases where AMD is unavoidable, it is necessary to treat it using an appropriate technique. Treatment technologies can be divided into passive and active treatment, both of which include biological, physical and chemical approaches. Active treatment requires continuous operation with regular addition of reactants and labor, while passive treatment requires only occasional maintenance [9].

(a)

(b)

Figure 1: Examples of effects of AMD in South Africa, showing the typical red/orange color due to iron hydroxide (pictures taken by the author).

Active Treatment

There are many active methods for the treatment of AMD, but the most common are: aeration, neutralization (including chemical precipitation), metal removal, chemical precipitation, membrane filtration, ion-exchange processes and biological sulfate removal [1].

Aeration

The objective of aeration is to oxidize dissolved Fe^{2+} as it is one of the main pollutants in AMD. If the wastewater contains more than 50 mg/l Fe^{2+} then it must be aerated. Aeration increases the level of dissolved oxygen (DO), promoting the oxidation of iron and manganese, which increases the efficiency of chemical treatment and thus reduces costs. During aeration, dissolved carbon dioxide from underground mine water will be released, resulting in an increase in the pH and a reduction in the cost of reagents [1].

Neutralization

AMD can be neutralized by chemicals such as sodium and calcium hydroxide and their carbonates in order to precipitate metals. Neutralization and precipitation are used quite often due to the feasibility of treating large volumes of contaminated water, the low cost and the simplicity of the process [10].

Quicklime (CaO) and hydrated lime (Ca (OH) $_2$) are used for the neutralization of AMD due to their abundance and high reactivity. During neutralization metals such as Fe^{2+}, Fe^{3+}, Al, Cu, Zn and Pb are precipitated in the form of metal hydroxides. The sludge resulting from this process is a mixture of metal hydroxides and gypsum ($CaSO_4$). Equation (6) below gives the main neutralization reaction when using hydrated lime [10].

$$Ca(OH)_2 + Me^{2+}/Me^{3+} + H_2SO_4 \Leftrightarrow Me(OH)_2/Me(OH)_3 + CaSO_4 + H_2O \quad (6)$$

Sludge containing Fe^{3+} is more stable than sludge containing Fe^{2+}, and air is therefore used during neutralization to oxidize Fe^{2+} to Fe^{3+}. Clarifiers or thickeners are used to settle the sludge produced, and if the solids content is less than 1 mg/l, sand filters can be used to polish the treated water. The solids content of sludge is strongly affected by the concentration of metals in the water and the type of treatment process applied, and can vary from 1% to 30%. The process is optimized by adjusting the process parameters (neutralization rate, oxidation rate, ratio of Fe^{2+}/Fe^{3+}, ion concentration, temperature, sludge age, crystal formation and sludge recycling) in order to obtain a denser sludge, thus reducing the volume [10] .

This process is called high sludge density (HSD) and it is a modification of conventional neutralization process and it aims to produce a higher sludge density [11]. Neutralization reactors are used to oxidize iron from Fe^{2+} to Fe^{3+} at certain pH. Treated water from the reactors is flocculated with a polymer, and the solids are separated from the liquid in a thickener or clarifier. The sludge produced in the thickener is routed back to the process [12]. The illustration of HSD process can be seen in Figure 2.

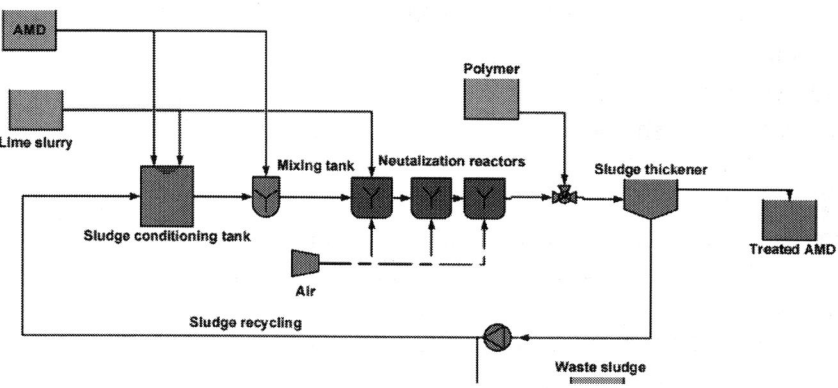

Figure 2: Basic configuration of the HDS [1].

The configuration illustrated in Figure 2 is the standard commercial HDS process used for the treatment of AMD, and has the following advantages: the low cost of lime and its efficient use, only a small site is required for sludge disposal due to high density of the sludge, the water/solid separation is good, and it is a very robust process, with the ability to treat AMD with different properties (flow, metal loading and acidity) [1].

Limestone has been used to treat AMD for many years in the coal mining industry because it is the cheapest material available, it is easy to handle, and is the safest chemical for treating AMD. The contaminants of greatest concern are iron and aluminum, and limestone is very effective in neutralizing these. However, the application of limestone is limited because it has a low solubility and has tendency to form an external coating of Fe $(OH)_3$ during the treatment of AMD [10] .

Under certain conditions, HDS can be achieved using limestone to neutralize AMD instead of lime [10]. Limestone reacts with acid water, leading to the dissociation and release of carbon dioxide, as in Equations (7) and (8) below.

$$CaCO_{3(S)} + H_2SO_{4(aq)} \Leftrightarrow CaSO_{4(S)} + H_2O + CO_{2(g)} \tag{7}$$

$$CaCO_{3(S)} + Fe_2(SO_4)_{3(aq)} + 3H_2O \Leftrightarrow 3CaSO_{4(S)} + 2Fe(OH)_{3(S)} + 3CO_{2(g)} \tag{8}$$

The carbon dioxide released forms carbonate ions, which buffer the pH to an upper limit of 6.5. As a consequence of this, some metals cannot be removed as they require a pH above 6.5 for precipitation. To overcome this problem, a combination of limestone and lime can be used, as shown in Figure 3.

This process has three different steps: 1) pre-neutralization with limestone, which is a little cheaper than lime; 2) neutralization with lime in order to reach a certain pH that is determined by the metal to be removed; and 3) adjustment of the pH and re-carbonation using carbon dioxide produced in the limestone neutralization reactor [1]. When choosing the appropriate neutralization agent for the treatment of acid water from a particular mining site, the following parameters must be take into account: the type of material (including transportability, storability and dosing), the hazardous properties of the material, the reliability and availability of suppliers of reactants, the efficiency of neutralization, problems such as coating, clogging and scaling of the equipment, and the cost of the process [10]. Neutralization and hydrolysis are key aspects in the treatment of AMD. Table 1 lists different types of alkalis and materials used to treat AMD [1].

To determine the amount of alkali required to treat a certain AMD, it is necessary to consider the cost of the alkali, the objective of the treatment (in this case, the removal of metals), and the effects of the residue produced (INAP, 2013). The data given in Table 1 can be used to determine the amount of alkali required to neutralize a certain amount of acid, and to estimate the cost of the alkali required to perform the task. The flowchart shown in Figure 4 and the data in Table 1 can be used to design an appropriate neutralization system for any coal mining wastewater, providing the flow rate and water chemistry of the AMD are known. In order to select the *Acidity is expressed as $CaCO_3$; **Market prices in January 2009.

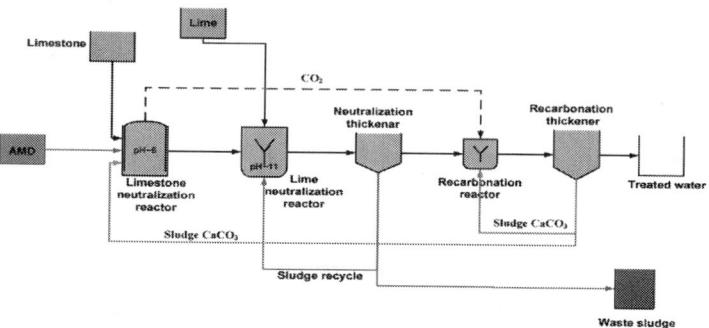

Figure 3: AMD treatment using a combination of limestone and lime [1].

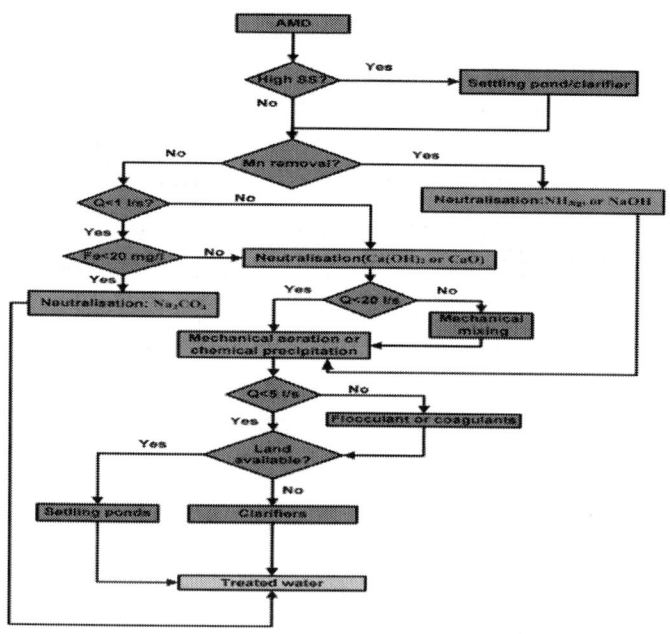

Figure 4: Flowchart that can be used to design a site-specific AMD treatment system [13].

Table 1: Materials and alkali applied for AMD treatment [1]

Neutralization agent (Alkali)	Dosage (ton of alkali/ton of acidity)*	Efficiency of neutralization (% of applied alkali)	Cost (USD/ton bulk)**
Limestone (CaCO$_3$)	1	30 - 50	10 - 15
Hydrated lime (Ca(OH)$_2$)	0.74	90	60 - 100
Un-hydrated lime (CaO)	0.56	90	80 - 240
Soda ash (Na$_2$CO$_3$)	1.06	60 - 80	200 - 350
Caustic soda (NaOH)	0.8	100	650 - 900
Magna lime (MgO)	0.4	90	Project-specific
Fly ash	Material-specific	-	Project-specific
Kiln dust	Material-specific	-	Project-specific
Slag	Material-specific	-	Project-specific

Appropriate neutralization agent, it is important to known the concentrations of iron and manganese. Manganese is very soluble in the pH interval 4.5 to 8 making its removal difficult. The best way to remove Mn is by raising the pH to a value above 9 in order to oxidize Mn^{2+} to Mn^{3+} or Mn^{4+}, allowing the insoluble manganese carbonate or manganese oxide to be removed [13] .

Passive Treatment

Various kinds of passive treatment can be used; the most common being aerobic wetlands, anaerobic wetlands, anoxic limestone drains, open limestone drains, and reducing and alkalinity-producing systems [1]. The critical parameters in the design of passive treatment systems for AMD are the flow, the properties of the AMD and land availability (Zipper, et al., 2011).

Aerobic Wetlands

The simplest type of passive treatment is the aerobic wetland, but it cannot be used to treat all types of acid water efficiently. Its capacity to neutralize acidity is limited, but it can be used to treat net alkaline water that has a high content of iron. Mine water is aerated while it flows slowly through the vegetation, and dissolved iron will thus be oxidized, and the oxidation product precipitated. The pH will fall as a result of the precipitation of iron due to the generation of H$^+$ ions, and the treated water will thus have a lower pH than the influent water, despite the fact that the iron concentration is higher in the influent water. Aerobic wetlands can also be used to remove Mn, but the oxidation of Mn only starts when the oxidation of Fe is completed. To remove Mn using aerobic wetlands it is necessary to have large areas to allow the complete oxidation of Fe and thus the oxidation of Mn. Alternatively, another wetland cell can be added. Figure 5 shows a typical aerobic wetland where aquatic plants transport oxygen through the roots to the subsurface to help the oxidation process. Composted organic matter or natural soil can be used as substrate, and water levels between 10 - 30 cm are used to maintain aerobic conditions, and to allow the cattails to grow in order to improve wetland performance [14].

Anaerobic Wetlands

Anaerobic wetlands are a modification kind of aerobic wetlands, where a bed of limestone and a layer of biodegradable organic matter are added in order to allow the treatment of acid water.

The limestone is located below the substrate to enhance the generation of alkalinity in the form of. HCO_3^- Under anoxic conditions (low oxygen levels) sulfate can be reduced in the presence of biodegradable organic matter. Sulfate-reducing bacteria use the oxygen in SO_4^{2-} that enters the system under anoxic conditions to reduce sulfate to H_2S gas or to a solid sulfide by the biodegradation of organic matter in a metabolic process [14].

This process is illustrated by Equation (9).

$$SO_4^{2-} + 2CH_2O \Rightarrow H_2S + 2HCO_3^-$$

(9)

If metals (M) are present in the solution, the reduction process leads to metal sulfides, as can be seen from Equation (10). These metal sulfides are deposited in the substrate.

$$M + SO_4^{2-} + CH_2O \Rightarrow MS + HCO_3^- \tag{10}$$

Alkalinity can also be generated by the reaction between acid water and the limestone below the substrate, as in Equation (11).

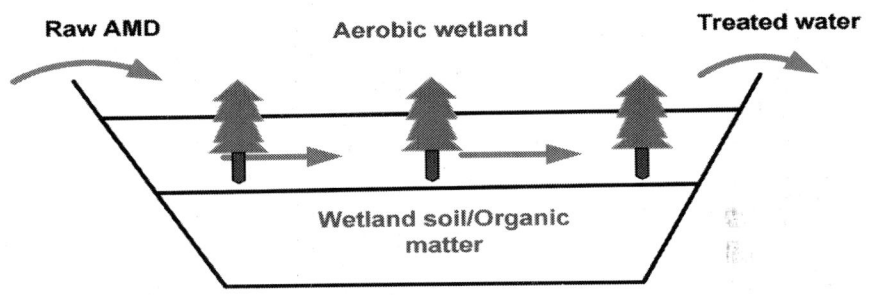

Figure 5: Cross section of aerobic wetland [14].

$$CaCO_3 + H^+ \Rightarrow Ca^{2+} + HCO_3^- \tag{11}$$

These three equations illustrate the production of bicarbonate ions, which are the source of alkalinity, and they can raise the pH by the neutralization of H^+ (Equation (12)), thus contributing to the precipitation of soluble metals present in acid water.

$$HCO_3^- + H^+ \Rightarrow H_2O + CO_2(aq) \tag{12}$$

Figure 6 illustrates anaerobic wetlands also known as composted wetlands and the curved arrows in Figure 6 indicate diffusion of water.

Anoxic Limestone Drains

Anoxic limestone drains (ALDs) (Figure 7) is an engineered method where limestone is used to intercept AMD in anoxic conditions. Limestone in contact with AMD dissolves and generates alkalinity. To avoid contact between oxygen and AMD, limestone is crushed and buried with compacted soil or clay and the effluent water is led to a settling pond where the pH is adjusted to bring about the precipitation of metals [15]. When ALDs are working properly, they are more cost-effective than wetlands, but they cannot be used to treat AMD with significant amounts of Fe^{3+}, Al and DO, due to clogging resulting from the precipitation of metal hydroxides when the pH reaches or exceeds 4.5. To avoid clogging, the influent concentrations of Fe^{3+}, Al, and dissolved oxygen must all be below 1 mg/l. Under anoxic condition armoring by iron hydroxide cannot take place because Fe^{2+} cannot precipitate as $Fe(OH)_2$ at a pH below 6 [14] .

Vertical Flow Systems

Most of AMD can be treated by the passive methods described above, but if mine water contains DO, Fe^{3+} and Al in great quantity, successive alkalinity-producing systems (SAPS) or vertical flow system which is a combination of ALDs and anaerobic wetlands with the objective of compensating for the limitations of each method can be used [16]. When AMD enters the system it flows vertically downwards through the organic layer where dissolved oxygen is removed by aerobic bacteria using biodegradable organic matter as their energy source, while other bacteria generate alkalinity by reducing sulfate to sulfide (Figure 8). The organic matter layer must be able to reduce the level of dissolved oxygen to less than 1 mg/l to avoid limestone armoring and to allow the reduction of sulfate. The limestone layer allows the dissolution of $CaCO_3$ by acid water, and under anoxic conditions more alkalinity will be produced. Finally, the water is discharged to a settling pond where the acid is neutralized and the metals precipitated. When the influent AMD contains a significant amount of Fe^{3+} and sediments, pretreatment in an aerobic wetland or a settling pond is necessary to avoid the accumulation of solids. When the influent is highly acidic, it is necessary to divide the system into several vertical flows that can be separated by different settling ponds [14].

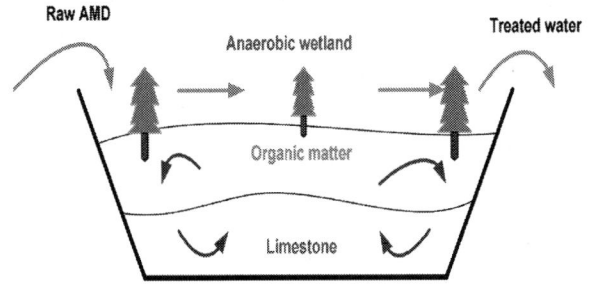

Figure 6: Cross section of an anaerobic wetland [14].

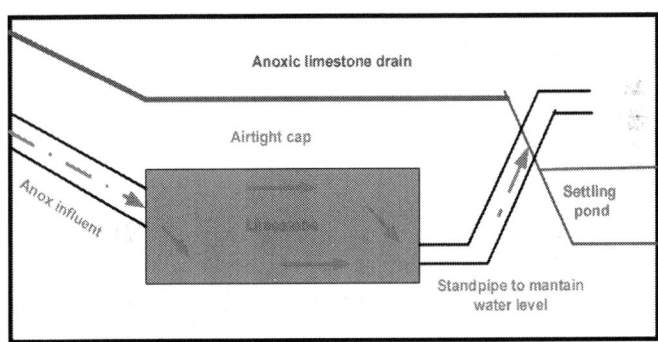

Figure 7: Cross section of an ALD system [14].

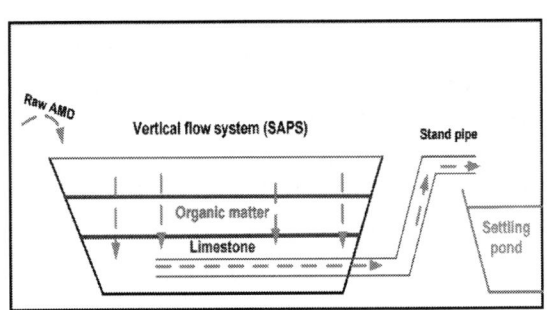

Figure 8: Cross section of a vertical flow system [14].

The flowchart in Figure 9 can be of use when choosing the appropriate kind of passive treatment of AMD. As with active methods, it is important to know the flow rate and the chemical composition of the AMD. Samples should be collected from tailing seepage or mine discharge, and the levels of Fe, Mn, alkalinity, pH and acidity measured (Hedin et al., 1994). The composition of AMD can change considerable with the seasons, and it is thus important to collect and analyze samples at different times of the year [17].

SUMMARY OF TREATMENT TECHNOLOGIES

Summaries of passive and active treatments are presented in Table 2 and Table 3, including the advantages and disadvantages of each method. These tables can be used to help decide which active or passive treatment is most suitable for a certain AMD.

To ensure successful treatment of the particular acid water, parameters such as acidity, flow rate, dissolved oxygen and pH should be analyzed. Table 4 can be used to decide which method can be used successfully in a particular case.

According to Skousen & Ziemkiewicz [20], it is necessary to take into account flow rate, water chemistry, topography and the characteristics of the area in order to select and design a suitable passive treatment system. Table 5 lists some aspects that should be considered during the selection and design of a passive treatment system for AMD.

Comparing Table 4 and Table 5 it can be seen that they give almost similar information. Combining these tables provides very useful information for the selection and design of a suitable passive treatment method.

SELECTION OF REMEDIAL TECHNIQUE

To select the appropriate remedial technique it is necessary to perform a feasibility study in each specific case. This process starts with the specification of the problem (e.g. chemicals, risks, etc.), followed by the identification of potential techniques and, finally, evaluation of the feasibility of the selected techniques. According to [21], the steps in evaluating the feasibility are, first: Effectiveness—"the potential for the alternative to achieve remedial goals established for the site", second: Implement ability—"the ability to comply with technical and administrative issues and constraints involved in implementing a technique at a specific site and third: Cost— "typically an estimate of net present cost for each technique". In practice, mining companies first identify the techniques that can meet their water quality goals (effectiveness), then they eliminate those that cannot be applied for practical reasons (implement ability), and finally, the least expensive method is implemented (cost) [23].

To select a suitable technique to treat a certain AMD, it is necessary to analyze the AMD and asses the available options based on the information given in Figure 4 and Figure 9. Table 4 and Table 5can also are used for construction and successfully treatment. Figure 10 summarizes what should be done to choose appropriate technique to treat a site specific AMD.

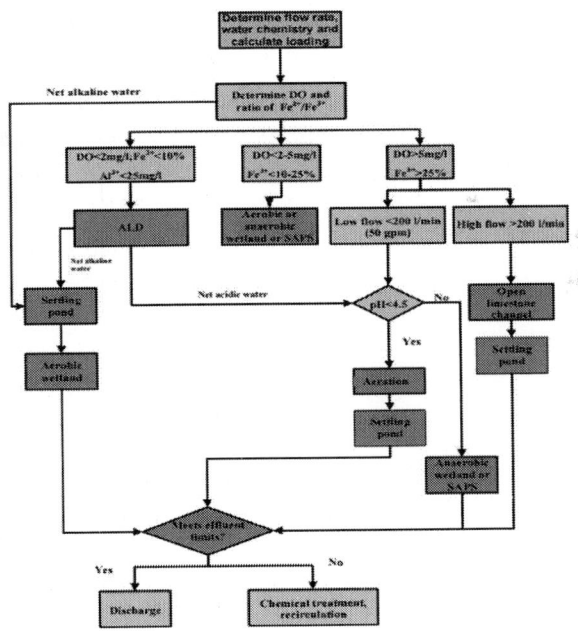

Figure 9: Flowchart to aid the selection of passive treatment method for AMD [18].

Table 2: Summary of passive methods of treating AMD

Treatment method	Suitable for	Advantages	Disadvantages	Reference
Open limestone channels	Pre-treatment or post-treatment	Low operating and maintenance costs, no power consumption, can last for many years, simple and reliable	Coating, long channel required to achieve the desired retention time	[13]
Aerobic wetlands	Acid water containing Fe, Mn and SS	Low operating and maintenance costs, no power consumption, can last for many years	Cannot treat strongly acidic water efficiently (best for pH > 5.5)	[17]
Anaerobic wetlands	Acid water with low DO, Al, Fe^{3+} and SS contents	Low operating and maintenance costs, no power consumption, can last for many years	Coating on limestone surface due to presence of Fe and Al, large area and long retention time needed to remove Mn	[17]
Anoxic limestone drains	Acid water with low Al and Fe^{3+} contents	Low operating and maintenance costs, no power consumption, can last for many years, simple	Needs pretreatment, best for acid water with low DO, Al and Fe^{3+} to avoid armoring	[14]
Successive alkalinity-producing systems	Acid water with high contents of metals (Fe, Al, Zn, Cu)	Low operating and maintenance costs, no power consumption, can last for many years	Metal floc accumulation and degradation of organic layer, pretreatment required for acid water with high Fe^{3+} and SS contents	[14]

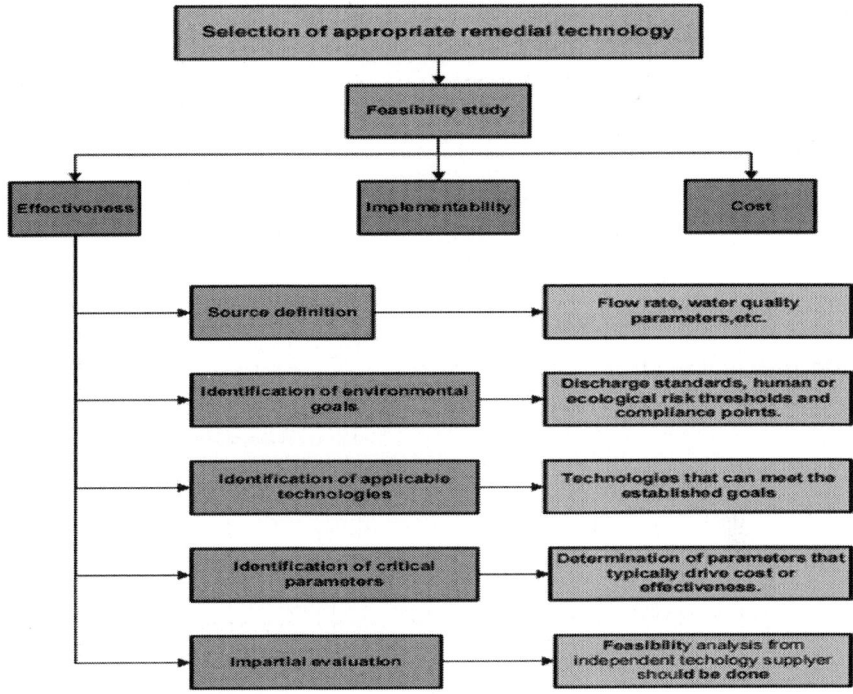

Figure 10: Method employed for the selection of the best technique for the treatment of AMD in specific cases.

Table 3: Summary of active methods of treating AMD

Treatment method	Suitable for removing	Advantages	Disadvantages	Reference
Aeration	Fe^{2+} and Mn	Low operation cost, releases CO_2 from mine water, increases DO	Not effective for water with low Fe^{2+} content	[19]
High-density sludge process	Fe^{2+}, Fe^{3+}, Al, Mn, Cu, Zn, Pb and SO_4^{2-}	Generation of low volumes of sludge, high water recovery, low lime cost, scaling control, can treat large flows of AMD, sludge recycling	Limited sulfate removal, generation of sludge	[10]

Limestone/ lime neutralization	Fe and Al	Low alkali cost, sludge recycling	Limited sulfate removal, generation of sludge	[20]
Membrane filtration	Brackish and saline mine water	Good quality of treated water, high water recovery	Scaling, fouling, needs pretreatment and post-treatment, sludge and brine production and short membrane life	[21]
Biological sulfate removal	Sulfate and Fe	Very effective in removing sulfates	Best for water with pH > 5, effluent metal concentration may exceed permissible limits	[22]

Table 4: Influent characteristics of AMD required for successful treatment [9]

Treatment method	Acidity range (mg $CaCO_3$/l)	Acidity load (kg $CaCO_3$/d)	Q (l/s)	DO (mg/l)	pH	Maximum pH attainable
Passive treatment						
Aerobic wetlands	<500	≤1	Maximum permissible residence time (e.g. 1 - 5 days)	Ambient	>6	-------
Anaerobic wetlands	<500	1	Maximum permissible residence time (e.g. 1 - 5 days)	Ambient near surface and <1 mg/l below surface	>2.5	6 - 8
ALD	<500	<150	< 20	<1	>2	6 - 8
SAPS	<300	<100	< 10	<1.3	>2.5	6 - 8
Active treatment						
All	1 - 10,000	1 - 50,000	No limit	------	No limit	14

Table 5: Influent AMD characteristics and design factors for successful passive treatment of AMD [20]

Treatment method	Requirements	Construction	Design factors
Ponds	Net alkaline water	None	None
Aerobic wetlands	Net alkaline water	Overland flow with cattails	10 - 20 g Fe/m²/day 0.5 - 1 g Mn/m²/day
Anaerobic wetlands	Net acidic water Low flow	Flow over and within the substrate	3.5 g acidity/m²/day
Sulfate-reducing bioreactors	Net acidic water low flow	Flow through the substrate	Residence time of 24 h
ALD	Net acidic water low DO, Al, Fe contents	Flow through buried limestone	Residence time of 15 h
SAPS	Net acidic water	Vertical flow	15 - 30 cm of organic matter, residence time of 15 h, 20 g acidity/m²/day
Open limestone drain	Slope > 10%	Rock-lined channel	Acid load and residence time
Limestone leach bed	Inflow pH < 3.0	Flow through limestone	Residence time of 1.5 h
Slag leach bed	Water without metals	Flow through steel slag fine aggregate	Residence time of 1 - 3 h

DISCUSSION AND CONCLUSIONS

The generation of AMD and its treatment are complex issues requiring careful analysis in each individual case. Prevention and containment of AMD are the best management strategies, as treatment is often costly. However, if prevention and containment are not possible, treatment must be applied to avoid contamination of the water resources surrounding mining site. The most appropriate technique for the treatment of AMD is site-specific, as it depends on the flow rate and chemistry of the acid water. Costs, implement ability and effectiveness must also be taken into consideration.

Many active and passive methods of treating AMD have been discussed to provide a general picture of the strategies applied by coal mining companies. Only the basic methods were described, whereas in reality there are many variations in use around the world. In general, active treatment methods are suitable in cases where no land is available and in cases where it is necessary to control the process. These methods can be used to remove pollutants from AMD efficiently, but the investment, maintenance and operating costs are high. Passive methods, on the other hand, are in general suitable for mines no longer in operation as they need less maintenance and operate naturally. However, they require large areas of land and long retention time to operate efficiently.

Methods of active treatment have several advantages, for example high removal efficiency, large volumes of AMD with different characteristics can be treated, the systems can be controlled automatically, and they occupy a relatively small area. However, they are associated with high costs and they generate sludge. The advantages of passive treatment systems are: low costs, they last for many years, and they do not require any power, but large areas and long retention times are required for them to operate efficiently.

In conclusion, active methods are more suitable for operating mines, while passive methods are suitable for closed mines.

ACKNOWLEDGEMENTS

First, I thank God for giving me strength, health and wisdom to write this paper. I also would like to express my gratitude to my supervisors, Lund University and Eduardo Mondlane University for all support that they gave me.

REFERENCES

1.   INAP (2013) The Global Acid Rock Drainage Guide. International Network for Acid Prevention (INAP), 2013.

2.   Lottermoser, B.G. (2010) Mine Wates: Characterization, Treatment and Environmental Impacts. 3rd Edition Edition, Queensland: Springer, 2010.

3.   Akcil, A. and Koldas, S. (2005) Acid Mine Drainage: Causes, Treatment and Case Studies. Cleaner Production, 2005.

4.   Skousen, J., Geidel, G., Foreman, R., Evans, R. and Hellier, W. (1998) A Handbook of Technologies for Avoidance and Remediation of Acid Mine Drainage. National Mine Land reclamation Center, Virginia.

5.   MEND (2008) Acid Rock Drainage Prediction Manual. Electronic Revision Edition, Mend, Pacoima.

6.   Jennings, S., Neuman, D. and Blicker, P. (2008) Acid Mine Drainage and Effects on Fish Health and Ecology: A Review. Reclamation Research Group, Alaska.

7.   Dharmappa, H., Wingrove, K., Sivakumar, M. and Singh, R. (1999) Wastewater and Storwater Minimisation in a Coal Mine. Journal of Cleaner Production, 8, 24-34.

8.   EPA (2008) Coal Mining Detailed Study. United States Environmental Protection Agency, Washington DC.

9.   Taylor, J., Pape, S. and Murphy, N. (2005) A Summary of Passive and Active Treatment Technology for Acid and Metalliferous Drainage. Earth Systems, Fremantle.

10. Kuyucak, N. (2006) Selecting Suitable Methods for Treating Mining Effluent. Golder Association.

11. Maree, J.P., Strydom, W.F., Adlem, C.J.L., de Beer, M., van Tonder, G.J. and van Dijk, B.J. (2004) Neutralization of Acid Mine Water and Sludge Disposal. CSIR.

12. DWA (2013) Feasibility Study for a Long Term Solution to Address the Acid Mine Drainage Associated with the East, Central and West Rand Underground Mining Basins. Department of Water Affairs (DWA), Pretoria.

13. Trumm, D. (2010) Selection of Active and Passive Treatment System for AMD—Flow Chart for New Zealand Conditions. Journal of Geology and Geophysics, 53, 195-210.

14. Zipper, C., Skousen, J. and Jage, C. (2011) Passive Treatment of Acid Mine Drainage. Virginia Tech, Blacksburg.

15. Watzlaf, G.R., Schroeder, K.T. and Kairies, C.L. (2000) Long Term Performance of Anoxic Limestone Drain. Mine Water and the Environment, 19, 98-110.

16. Ordóñez, A., Loredo, J. and Pendás, F. (2012) A Successive Alkalinity Producing System (SAPS) as Operational Unit in a Hybrid Passive Treatment System for Acid Mine Drainage. Mine Water and Environment, 575-580.

17. Hedin, R., Narin, R. and Kleinmann, R. (1994) Passive Treatment of Coal mine Drainage. Bureau of Mines.

18. Ford, K. (2003) Passive Treatment Systems for Acid Mine Drainage. Technical Note 409. Bureau of Land Management, Colorado.

19. Magdziorz, A. and Sewerynsky, J. (2000) The Use of Membrane Technique in Mineralize Water Treatment for Drinking and Domestic Purposes at Pakoj Coal Mine District Under Liquidation. Central Mining Institute, Department of Water Protection.

20. Skousen, J. and Ziemkiewicz, P. (2005) Performance of 116 Passive Treatment Systems for Acid Mine Drainage. Proceedings of the 2005 National Meeting of the American Society of Mining and Reclamation, Breckenridge, 19-23 June 2005, 1103.

21. EPA (2006) Management and Treatment of Water from Hard Rock Mine. United States Environmental Protection Agency, Washington DC.

22. Kirby, C., Dennis, A. and Kahler, A. (2009) Aeration to Degas CO_2, Increase pH and Increase Iron Oxidation Rates for Efficient Treatment of Alkaline Mine Drainage. Applied Geochemistry, 24, 1175-1184.

23. Geldenhuys, A., Maree, J., Beer, M. and Hlabela, P. (2003) An Integrated Limestone/Lime Process for Partial Sulphate Removal. The Journal of the South African Institute of Mining and Metallurgy, 345-354.

Spatio-temporal Land Cover Dynamics in Open Cast Coal Mine Area of Singrauli, M.P., India

Imran Khan and Akram Javed

Department of Geology, Aligarh Muslim University, Aligarh, India

ABSTRACT

Singrauli is an opencast coal mining area where large scale mining activities are going on continuously, land use/land cover studies are of vital importance to observe the changes in the land use/land cover. The present study utilizes multi-spectral/multi-temporal data of Indian Remote Sensing Satellite (IRS) LISS II geocoded (FCC) of 4th May 1993 and LISS III of 4th May 2010 for thematic mapping. Survey of India toposheet 63L/12 on scale 1:50,000 were used for preparation of base map which was overlaid on the FCC for land use/land cover mapping through visual interpretation. Visual interpretation of satellite imagery led to the identification of 15 land use/land cover categories such as dense forest, open forest, open scrub, plantation, cultivated land, uncultivated land, mining pit, overburden dumps, wasteland and settlement. The ground truth verification was carried out in key areas to rectify the errors in generated maps and then land use/land cover maps were finalized. The comparative analysis of land use/land cover shows that dense forest has been degraded to open forest, open scrubs and mining pits due to the expansion of mining activity. Open scrubs has increased, overburden dumps has increased, settlement has also increased, cultivated land has decreased and changed into uncultivated land and wasteland. It has also been observed that the plantation has been done on overburden dumps and residential colonies of NCL and NTPC. It has been identified that the main drivers which has increased the rate of change in land use/land cover are mainly coal mining activities and industrial expansion.

INTRODUCTION

Coal mining is one of the core industries that contribute to the economic development of India but deteriorates the environment [1]. The industrial expansion requires enormous energy generation for which huge amount of coal is extracted through mining which causes widespread landscape destruction [2]. As for environmental point of view coal mining is a major habitat transforming activity which has a number of adverse environmental consequences, namely soil erosion, acid mine drainage and increased sediment load as a result of abandoned and un-reclaimed mined land [3]. Land, air and water are adversely affected by different stages of mining processes and results in serious environmental degradation [4,5].

Land use/land cover changes has been described as the most significant regional anthropogenic disturbance to the environment [6], and are consistently associated with mining of natural resources [7]. Detailed knowledge of land use practice, land use pattern changes with time and its effects on environment and system are important to understand the importance of changes in land use [8]. Mining, especially open cast mining has a major impact on landscape, during pre-mining and post mining operation. Large scale coal mining operations have considerable impact on the pre-mining environmental scenario [9]. Removal of vegetation cover results in soil degradation due to accelerated water erosion, soil compaction and soil crusting, affecting productivity of land [10]. Mining activities also disturbs large tract of land due to overburden dumps which change the natural topography and drainage pattern of the area [9]. The change in land cover due to the coal mining activities particularly deforestation, has attracted worldwide attention because of their potential effects on soil erosion, run off and carbon dioxide level. Large scale deforestation has been reported in India in the past [11].

The concept, method and application of land use/land cover studies are introduced to mining area in order to find the land use change and give support to land management and ecological reconstruction [12]. It's prerequisite for planning, policy making and developmental programs that land use/land cover information its spatial distribution and change in land use pattern is commonly used. Information about land use/land cover not only provides better understanding of land utilization aspects but also plays important role in developmental planning [13]. The studies of land use/land cover of coal mining area will help in understanding the mechanism of land use/land cover, the economic activities of human society would be adjusted and make the land use more reasonable so as to achieve the purpose of land resource continual use [14].

Remote sensing and Geographic information system (GSI) are important tools for studying the land use pattern and their dynamics. The mapping of land use classes and monitoring their changes with time has been widely recognized [7]. The change detection in land use/land cover due to natural and human activities can be monitored by using multidate images to evaluate differences in land cover [15]. The present study makes an attempt to quantify land cover changes in Singrauli industrial belt using multi temporal remote sensing data, supported by topographic map, Census of India reports, revenue records and ground truth data. The main objective of present study is to understand land use/land cover change in time and space, with special references to the coal mining activities.

STUDY AREA

The study area lies partly in Singrauli district of Madhya Pradesh and partly in Sonebhadra district of Uttar Pradesh, bounded within the geo-coordinates 24°00›N to 24°15›N latitudes and 82°30›E to 82°45›E longitudes, (Figure 1). The total area of the coal field is 2201 km², however at present only 300 km² area is exploited for coal. With the availability of power grade coal reserves and the nearby water reservoirs (Govind Ballabh Pant Sagar) offers an excellent location for super thermal power plants (STPS), Aluminium plants, and cement industries. Singrauli coalfield has coal reserves of 1789.41 million tonnes with a steep rise in coal production from 30.70 million tonnes in 1993 to 67.67 million tonnes in 2010. The area is well suited for thermal power generation and is expected to produce 25% of thermal power need of India [16]. The climate of the district Singrauli is tropical monsoonal dry during the period of November up to June while during rainy season the atmosphere is very humid. The average rainfall received during the last 26 years is 1177.06 mm, the maximum and the minimum rainfall received is 1457.30 mm and 798.60 mm in 1978 & 1979, respectively. The geological sequence of the area is represented by series of formations that have been recognized on broad lithic characteristics within the Gondwana rocks of Singrauli coal field. In ascending order they are Talchir, Barakar, Barren Measure, Raniganj, and Mahadeva. However only four Precambrian, Talchir, Barakar, Barren Measure has been reported from study area. The most dominant geological formation in the area is Barakar formation covering 515.18 km² followed by Talchir covering 118.94 km², Precambrian covering 57.63 km² and Barren Measure covering 16.89 km² respectively. The area is occupied by structural hills on the northern part with elevation ranging between 270 m - 620 m, formed of resistant Precambrian rocks. The structural plateau is made up of coalbearing Gondwana rocks. Low lying flats characterized by gentle-undulating topography in the central part of the area where most of the agricultural activities have been noticed [17]. The Kachni, Mayar, Motwani and Baliya nala are four main dendritic perennial streams or river transverse through the Singrauli coal field which are mostly seasonal.

The development of the Singrauli area began during the year 1950s with the construction of two dams on the Rihand River. These reservoirs were mainly for irrigation purpose but 400 MW of hydropower has also been generated. Before the Coal mining and other industrial activities the region was densely covered with tropical deciduous forests. Coal mining operation on large scale has significantly changed the pre-mining environment scenario. In addition to mining activity, the surrounding industrialization have also an adverse impact on land use/ land cover, air quality and water quality of the study area.

DATA USED AND METHODOLOGY

Indian Remote Sensing Satellite (IRS) LISS II geocoded False Colour Composite (FCC) data of 4th May 1993 and IRS LISS III (FCC) of 4th May 2010 were used to analyze the land use/land cover pattern. Data available of the same season gives uniform spectral and radiometric characteristics and minimize the seasonal variation. The survey of India topographic sheets No 63L/12 scale 1:50,000 of 1976 was used for preparation of base map. The climate data of district Singrauli from 1978-2010 has been obtained from the Indian Meteorological Department (IMD) Pune. Secondary data obtained from published and unpublished sources such as internet (www.ncl.nic.in, www.ntpc.com, www.singrauli.nic.in) and district statistical handbook have been also used.

Visual interpretation is the effective method for classifying land use/land cover especially when the analyst is familiar with the area being classified from satellite data. This method uses skill that were originally developed for interpreting aerial photographs, and takes into consideration various photographic and geotechnical elements such as tone, texture, shape, size, association, drainage, landform and relationship with other objects to identify different land cover classes [18]. Land use/land cover change information can be obtained by either image-to image comparison or map-to-map comparison [19]. Using map-to-map comparison, images are to be classified and then maps are generated to compare which gives complete detail of land use/land cover changes.

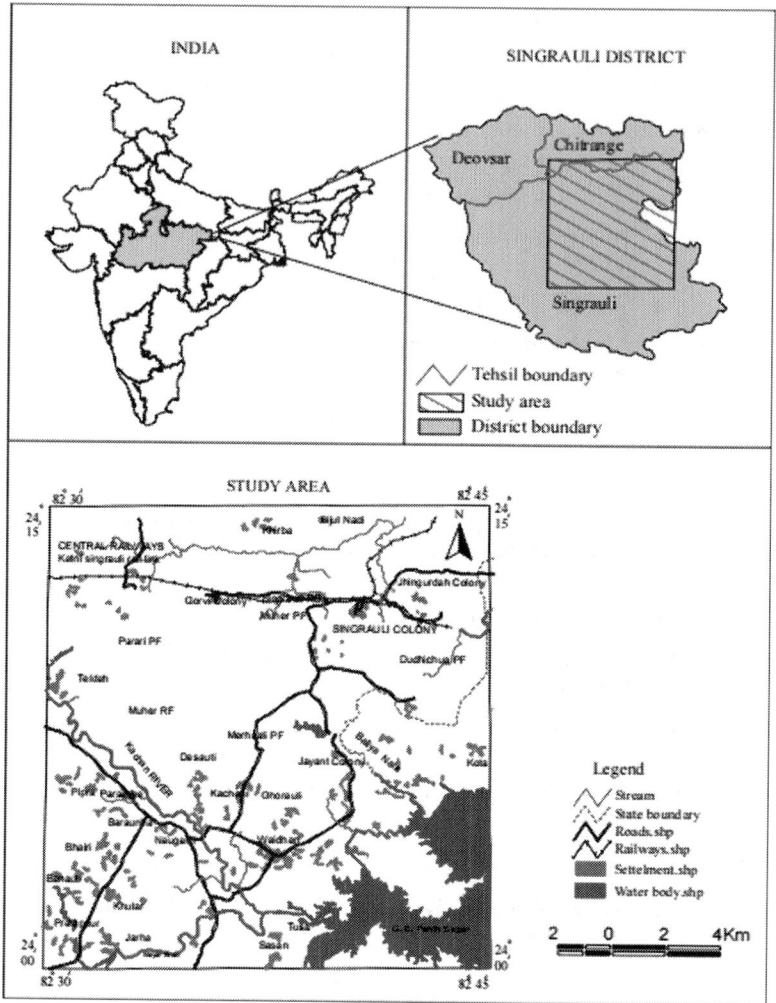

Figure 1: Location map of the study area.

In the present study, map-to-map comparison was used for land use/ land cover change detection. The images of both the years (1993 and 2010) were registered with topographical map to minimize geometric errors. Base map of the area having details such as settlement, road, railway line network, rivers and water bodies etc were superimposed on geocoded False Colour Composite (FCC) data for visual interpretation which led to the identification of various land cover categories in the study area. Land cover were categorized in 15 classes namely dense forest, open forest, open scrub, plantation, cultivated land, uncultivated land, mining pit, overburden dumps, waste land, rocky area, settlement, ash pond, water body, thermal power plant and dry river channel. These categories were identified on the basis of visual interpretation of satellite data and ground truth verification were done in the key areas for editing and authentication. On screen digitization technique has been carried out to digitize the maps using ArcMap 10 software for land use analysis. The steps followed for analysis are a) Digitization of land use map; b) Creation of polygon topology assigning unique id for each polygon; c) Editing. Area statistics of land use categories have been calculated in ArcMap 10 in sq.km as well as in percentage. The change in the extent of different land use categories during the period from 1993-2010 was analyzed and computed.

RESULTS AND DISCUSSION

Significant land use/land cover changes have taken place during 1993-2010 period due to natural and anthropogenic activities. Land use/land cover information derived from IRS LISS II 1993 (Figure 2), and IRS LISS III 2010 (Figure 3). Area statistics of each land use/land cover category of 1993 and 2010 were generated in ArcMap 10 and has been determined to analyze change in their spatial distribution (Table 1). By comparing the land use/land cover maps of 1993 and 2010 a change detection map has been automatically generated in Arc View GIS software to assess the major changes in the land use/land cover during 1993-2010 (Figure 4). The comparative analysis of each land use/land cover statistics reveals that, the dense and open forest are the important land cover categories which together constitute 273 km^2, i.e. 37.5% of the total area. The rest 62.5% area comprise open scrub, plantation, cultivated land, uncultivated land, mining pit, overburden dumps, wasteland, rocky area, settlement, ash pond, water bodies, thermal power plant and dry river.

Land use/land cover categories which shows increase in area are, open forest occupies an area of 144.0 km^2 in 1993 and 148.09 km^2 in 2010, shows an increase of 4.1 km^2 in 2010 in the northern, north-western and northeastern side at an elevation of 344 m to 501 m whereas on southern side elevation ranges from 265 m - 343 m with gentle slope. The mining pits has increased from 7.5 km^2 in 1993 to

11.12 km² in 2010, shows total increase of 3.6 km² area, whereas, overburden dumps has also increased from 18.6 km² in 1993 to 39.20 km² in 2010 showing an increase of 20.6 km² area respectively. Mining pits and overburden dumps are mostly associated with high elevation ranging from 400 m to 550 m in the central part and slope vary from moderate to very steep, whereas towards the north western side Gorvi mine is categorized under gentle slope with higher elevations (475 m to 550 m). Open scrub shows an increase of 17.7 km² from 24.6 km² in 1993 to 42.24 km² in 2010. Open scrub occupies most of the area in southern side at an elevation of 344 m to 421 m with gentle slope. Plantation area shows increase of 12.6 km² from 10.8 km² to 23.43 km² in 2010 in the central part near the mining/industrial activities at an elevation of 400 m to 500 m with gentle to moderate slope. The uncultivated land occupies 142.6 km² in 1993 which has rose to 147.5 km² in 2010, registering an increase of 4.9 km², wasteland shows increase of 6.3 km²from 24.8 km² in 1993 to 31.05 km² in 2010. The settlement area which includes rural settlement, urban settlement and Industrial residential complexes shows increase of 12.8 km² from 32.2 km² in 1993 to 44.97 km² in 2010. The ash pond increases from 2.3 km² in 1993 to 8.44 km² in 2010, showing increase of 6.2 km². The dry river also shows increase of 2.9 km² from 1.7 km² in 1993 to 4.60 km² in 2010.

The land use/land cover categories which show decrease in area from 1993 to 2010 are dense forest occupies 129.0 km² in 1993 and 80.10 km² in 2010, showing a decrease of 48.9 km² which is mostly associated with higher elevation ranging from 442 m to 618 m in the north, north-east and north-west with moderate to very steep slopes. Cultivated land show decreases of 21.2 km² from 113.0 km² in 1993 to 91.80 km² in 2010 and water body also show decrease of 21.6 km² from 59.1 km² in 1993 to 37.50 km² in 2010. Cultivated land and uncultivated land mostly occupy the area in the north, north-east and north-west at an elevation of 344 m to 421 m elevation however in southern side elevation ranges from 265 m to 304 m elevation and falls under gentle slope.

The change analysis shows decrease in dense forest is primarily due to the rapid expansion of coal mining activity in dense forest area which has resulted in clearance of forest near mining area. But at places it has been converted to open forest due to reduction in canopy cover. The decline in the area under cultivated land is attributed to decline in rainfall which adversely affects the rainfed agriculture as has been corroborated by rainfall data analysis. Water table decreases from ground level also support the finding as sufficient water is not available for irrigation to support agriculture.

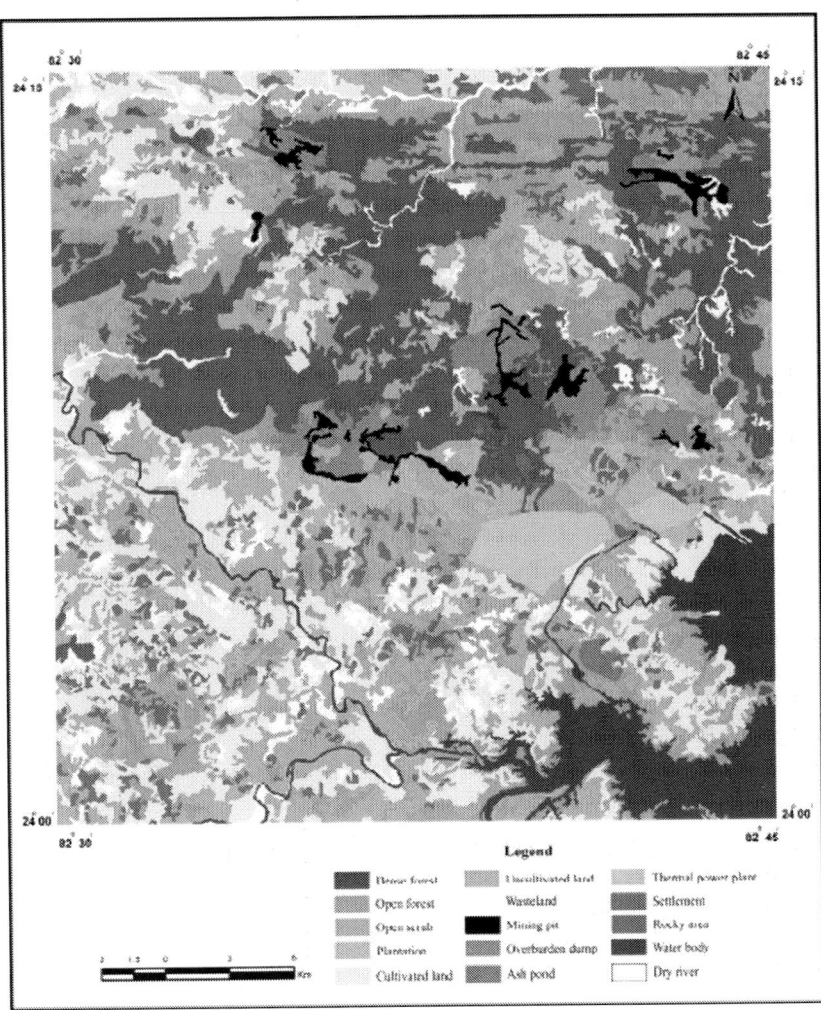

Figure 2: Land use/land cover map of Singrauli area based on IRS LISS II (1993).

Due to expansion of industries and coal mining farmers have been relocated resulting in loss of cultivated land.

The open forest has been increased because dense forest has degraded and converted to open forest by removing large number of trees for rapid expansion of mining activities in the forest area.

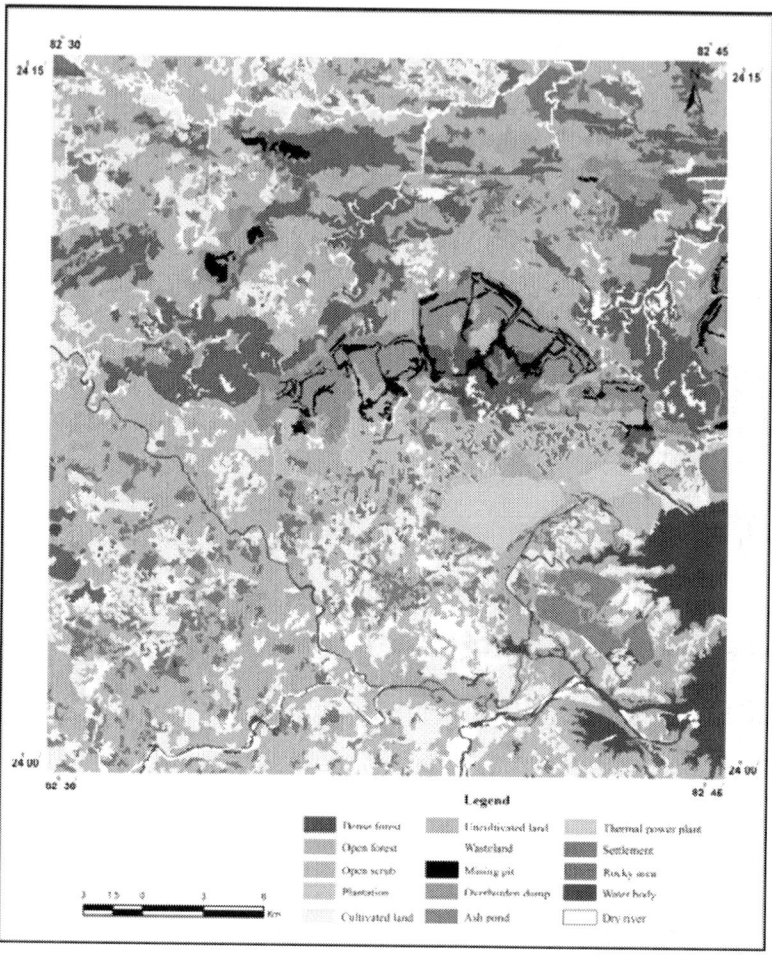

Figure 3: Land use/land cover map of Singrauli area based on IRS P6 LISS III (2010).

The increase in area is probably due to the degradation of dense forest into open forest near Parari and Mehrauli protected forest, Teldha and Jhingurdah mines.

Table 1: Details of the land use/land cover changes in the study area during 1993-2010

Land Use Categories	Land Use/Land Cover (1993)		Land Use/Land Cover (2010)		1993-2010 Net Change (sq kin)	1993-2010 Net Change (%)
	Area in (km²)	Area in (%)	Area in (km²)	Area in (%)		
Dense Forest	129.0	17.7	80.10	11.00	-48.9	-6.7
Open Forest	144.0	19.8	148.09	20.33	4.1	0.6
Open Scrubs	24.6	3.4	42.24	5.80	17.7	2.4
Cultivated Land	113.0	15.5	91.80	12.60	-21.2	-2.9
Uncultivated Land	142.6	19.6	147.50	20.25	4.9	0.7
Mining Pit	7.5	1.0	11.12	1.53	3.6	0.5
Overburden Dumps	18.6	2.6	39.20	5.38	20.6	2.8
Waste Land	24.8	3.4	31.05	4.26	6.3	0.9
Rocky Area	7.6	1.0	7.60	1.04	0.0	0.0
Settlement	32.2	4.4	44.97	6.17	12.8	1.8
Ash Pond	2.3	0.3	8.44	1.16	6.2	0.8
Water Body	59.1	8.1	37.50	5.15	21.6	-3.0
Thermal Power Plant	10.8	1.5	10.80	1.48	0.0	0.0
Dry River	1.7	0.2	4.60	0.63	2.9	0.4
Plantation	10.8	1.5	23.43	3.22	12.6	1.7
Total	728.4	100.0	728.4	100.0	183.2	25.98

Open forest has also come up at the periphery of the G. B. Pant sagar reservoir due to drying up of water from 1993 to 2010 where natural vegetation has grown near Tusa/Sasan and Devri. Plantation increase due to the reclamation of overburden dumps under operation "Green Gold" launched by Northern Coalfield Ltd and National Thermal

Power Cooperation has also carried out plantation programs around the colonies under social forestry scheme. The decline of water body by 21.6 km^2 area is due to the decrease in annual rainfall, siltation caused by the material run off from dump sites and huge amount of water is utilized for industrial purpose like cooling of generators, coal washery, at coal dumps to stop dust generation. The increase in the area of overburden dumps is due to the huge removal of material from mining blocks which were dumped along the edges of the plains and form artificial landscape, which constitute 5.38% of the total area in 2010. It was estimated that the overburden dump removal increases from 100.64 M cum in 1995 to 182.1 M cum in 2010. The increase in the area of ash pond is due to the increasing power generation capacity from super thermal power station which results burning of huge quantities of coal. The fly ash generated from the thermal power plants is disposed through the pipelines in the ash pond along the boundary of the G. B. Pant Sagar Reservoir. On the basis of 1993 and 2010 satellite data analysis, it is found that 183.2 km^2 (25.98%) area has changed from one land use/land cover to another land use/land cover category (Table 1).

CONCLUSIONS

The present study reveal that mining and industrial activities around Singrauli coal field are the main forces responsible for land use/land cover change during (1993- 2010). The mining area has increased tremendously that has resulted in degradation of dense forest, cultivated land and water bodies in the area. Due to deforestation, dense forest in the nearby mining area degrades and converts to open forest and open scrubs which are increasing in aerial extent whereas plantations at overburden dumps under reclamation schemes has been done. Settlement area has increases due to population explosion and peoples from other states come to this area for employment in mining and industrial sector. Degradation of natural resource in the area is going on due to the expansion of coal mining activities, increase in thermal power plants from two to six and other industrial activities. To prevent the enormous and uncontrolled loss of natural resources in this region, land use planning for degraded land and waste land should be implemented soon to minimize the impact. It may be concluded that the land use/land cover change in the Singrauli coal field has taken place due to the rapid expansion of mining and industrial activity during the period 1993 to 2001.

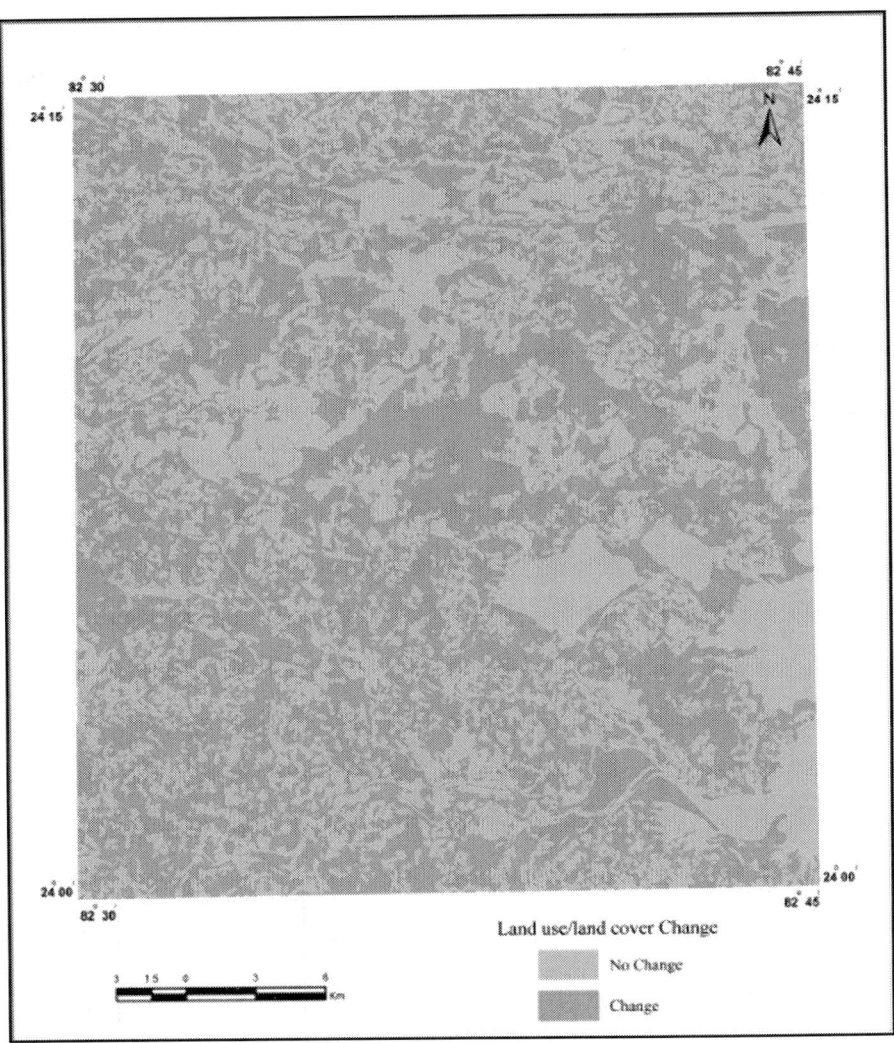

Figure 4: Change detection map (1993-2010).

This has resulted in the drastic changes in the land cover dynamics of the fragile ecosystem.

ACKNOWLEDGEMENTS

The authors are indebted to the Chairman, Department of Geology, AMU, Aligarh, for co-operation and providing necessary infrastructural facilities. Author is also grateful to University Grants Commission (UGC) for providing financial assistance under Non-NET fellowship to Ph.D. students. NRSA, Hyderabad and IMD, Pune are thankfully acknowledged for providing the data.

REFERENCES

1. R. K. Tiwary, "Environmental Impact of Coal Mining on Water Regime and Its Management," Water, Air, and Soil Pollution, Vol. 132, No. 1-2, 2001, pp. 185-199. doi:10.1023/A:1012083519667

2. A. N. Singh and J. S. Singh, "Experiments on Ecological Restoration of Coal Mine Spoil Using Native Trees in a Dry Tropical Environment, India: A Synthesis," New Forests, Vol. 31, No. 1, 2006, pp. 25-39.

3. N. F. Parks, G. W. Peterson and G. M. Baumer, "High Resolution Remote Sensing of Spatially and Spectrally Complex Coal Surface Mines of Central Pennsylvania: A Comparison between SPOT, MSS and Landsat-TM," Photogrammetric Engineering and Remote Sensing, Vol. 53, No. 4, 1987, pp. 415-420.

4. C. G. Down and J. Stocks, "The Environmental Impact of Mining," Applied Science Publisher, London, 1977.

5. F. G. Bell, S. E. T. Bullock, T. F. J. Halbich and P. Lindsey, "Environmental Impact Associated with an Abandoned Mine in the Witbank Coalfield, South Africa," International Journal of Coal Geology, Vol. 45, No. 2-3, 2001, pp. 195-216. doi:10.1016/S0166-5162(00)00033-1

6. D. A. Robert, G. T. Bastitsa, S. L. G. Pereira, E. K. Walter and B. W. Nelson, "Change identification Using Multitemporal Spectral Mixture Analysis—Application in Eastern Amazonian," In: R. S. Lunetta and C. D. Alvidge, Eds., Remote Sensing and Change Detection Environmental Monitoring Methods and Application, Sleeping Bear Press Inc., Arendal, 1998.

7.   A. Prakash and R. P. Gupta, "Land Use Mapping and Change Detection in a Coal Mining Area—A Case Study in the Jihar Coalfield, India," International Journal of Remote Sensing, Vol. 19, No. 3, 1998, pp. 391-410. doi:10.1080/014311698216053

8.   V. M. Varghese, B. Rajan, A. P. Pradeepkumar and R. Stephen, "GIS-Based Land Use/Land Covers Change Characterization in the Humid Tropical Meenachil River Basin, Kerala South India," Conference on Applied Geoinformatics for Society and Environment (AGSE), Arequipa, 3-6 August 2010, 6 pp.

9.   B.B. Dhar, A. Jamal and S. Ratan, "Air Pollution Problem in an Indian Open Cast Coal Mining Complex: A Case Study," International Journal of Surface Mining, Reclamation and Environment, Vol. 5, No. 2, 1991, pp. 83-88. doi:10.1080/09208119108944290

10. A. Manu, Y. A. Twumasi and T. L. Coleman, "Application of Remote Sensing and GIS Technologies to Assess the Impact of Surface Mining at Tarkwa, Ghana," IEEE International IGARSS Proceedings of Geosciences and Remote Sensing Symposium, Anchorage, 20-24 September 2004.

11. NRSA, "Manual of Nationwide Land Use/Land Cover Mapping Using Satellite Imagery," National Remote Sensing Agency, Hyderabad, 1989.

12. P. Du, H. Zhang, P. Liu, K. Tan and Z. Yin, "Land Use/Land Cover Change in Mining Areas Using MultiSource Remotely Sensed Imagery," International Workshop on Analysis of Multi-Temporal Remote Sensing Images, Leuven, 18-20 July 2007, pp. 1-7. doi:10.1109/MULTITEMP.2007.4293074

13. P. S. Dhinwa, S. K. Pathan, S. V. C. Sastry, M. Rao, K. L. Majumder, M. L. Chotani, J. P. Singh and R. L. P. Sinha, "Land Use Change Analysis of Bharatpur District Using GIS," Journal of the Indian Society of Remote Sensing, Vol. 20, No. 4, 1992, pp. 237-250. doi:10.1007/BF03001921

14. W.-B. Wu, J. Yao and T.-J. Kang, "Study on Land Use Change of the Coal Mining Area Based on TM Image," Journal of Coal Science and Engineering, Vol. 14, No. 2, 2008, pp. 287-290. doi:10.1007/s12404-008-0062-9

15. A. Singh, "Review Article—Digital Change Detection Techniques Using Remote Sensed Data," International Journal

of Remote Sensing, Vol. 10, No. 6, 1989, pp. 989-1003. doi:10.1080/01431168908903939

16. A. Jamal, B. B. Dhar and S. Ratan, "Acid Mine Drainage Control in an Opencast Coal Mine," Mine Water and the Environment, Vol. 10, No. 1, 1991, pp. 1-16.

17. S. Majumder and K. Sarkar, "Impact of Mining and Related Activities on Physical and Cultural Environment of Singurali Coalfield—A Case Study through Application of Remote Sensing Techniques," International Journal of Remote Sensing, Vol. 22, No. 1, 1994, pp. 45-56. doi:10.1007/BF03015119

18. T. M. Lillesand and R. W. Kiefer, "Remote Sensing and Image Interpretation," 5th Edition, John Wiley, New York, 2004.

19. K. Green, D. Kempka and L. Lackley, "Using Remote Sensing to Detect and Monitor Land Cover and Land Use Changes," Photogrammetric Engineering and Remote Sensing, Vol. 60, No. 3, 1994, pp. 331-337.

Citations

CHAPTER 1

Zaobao Liu, Jianfu Shao, Weiya Xu, and Chong Shi, "Estimation of Elasticity of Porous Rock Based on Mineral Composition and Microstructure," Advances in Materials Science and Engineering, vol. 2013, Article ID 512727, 10 pages, 2013. doi:10.1155/2013/512727.

CHAPTER 2

Agnieszka Dudzińska, "The Effect of Pore Volume of Hard Coals on Their Susceptibility to Spontaneous Combustion," Journal of Chemistry, vol. 2014, Article ID 393819, 7 pages, 2014 doi:10.1155/2014/393819.

CHAPTER 3

M. Garcia, O. Betancourt, E. Cueva and J. Gimaraes, "Mining and Seasonal Variation of the Metals Concentration in the Puyango River Basin—Ecuador," *Journal of Environmental Protection*, Vol. 3 No. 11, 2012, pp. 1542-1550. doi: 10.4236/jep.2012.311170.

CHAPTER 4

Al-Ruzouq, R. and Rawashdeh, S. (2014) Geomatics for Rehabilitation of Mining Area in Mahis, Jordan. Journal of Geographic Information System, 6, 123-134. doi: 10.4236/jgis.2014.62014.

CHAPTER 5

Q. Yu, H. Shimada, T. Sasaoka and K. Matsui, "Impact of Underground Mining on Shaft Lining and Aquifer in Eastern Chin," *Open Journal of Geology*, Vol. 2 No. 3, 2012, pp. 158-164. doi: 10.4236/ojg.2012.23016.

CHAPTER 6

Idris, G., Asuen, G. and Ogundele, O. (2014) Environmental Impact on Surface and Ground Water Pollution from Mining Activities in Ikpeshi, Edo State, Nigeria. *International Journal of Geosciences*, 5, 749-755. doi:10.4236/ijg.2014.57067.

CHAPTER 7

M. Zhang, H. Shimada, T. Sasaoka and K. Matsui, "Lateral Stress Concentration in Localized Interlayer Rock Stratum and the Impact on Deep Multi-Seam Coal Mining," *International Journal of Geosciences*, Vol. 4 No. 9, 2013, pp. 1248-1255. doi: 10.4236/ijg.2013.49119.

CHAPTER 8

K. Roach, N. Jacobsen, C. Fiorello, A. Stronza and K. Winemiller, "Gold Mining and Mercury Bioaccumulation in a Floodplain Lake and Main Channel of the Tambopata River, Perú," Journal of Environmental Protection, Vol. 4 No. 1, 2013, pp. 51-60. doi: 10.4236/jep.2013.41005.

CHAPTER 9

W. Pereira, A. Kelecom and J. Pereira, "Analysis of Radium Isotopes in Surface Waters nearby a Phosphate Mining with NORM at Santa Quitéria, Brazil," Journal of Environmental Protection, Vol. 5 No. 3, 2014, pp. 193-199. doi: 10.4236/jep.2014.53023.

CHAPTER 10

Pondja Jr., E., Persson, K. and Matsinhe, N. (2014) A Survey of Experience Gained from the Treatment of Coal Mine Wastewater. *Journal of Water Resource and Protection*, **6**, 1646-1658. doi: 10.4236/jwarp.2014.618148.

CHAPTER 11

I. Khan and A. Javed, "Spatio-Temporal Land Cover Dynamics in Open Cast Coal Mine Area of Singrauli, M.P., India," Journal of Geographic Information System, Vol. 4 No. 6, 2012, pp. 521-529. doi: 10.4236/ jgis.2012.46057.

Index